Six-Minute Solutions
for Civil PE Exam Environmental Problems

R. Wane Schneiter, PhD, PE, DEE

Professional Publications, Inc. • Belmont, CA

How to Locate Errata and Other Updates for This Book

At Professional Publications, we do our best to bring you error-free books. But when errors do occur, we want to make sure that you know about them so they cause as little confusion as possible.

A current list of known errata and other updates for this book is available on the PPI website at **www.ppi2pass.com**. From the website home page, click on "Errata." We update the errata page as often as necessary, so check in regularly. You will also find instructions for submitting suspected errata. We are grateful to every reader who takes the time to help us improve the quality of our books by pointing out an error.

SIX-MINUTE SOLUTIONS FOR CIVIL PE EXAM ENVIRONMENTAL PROBLEMS

Current printing of this edition: 2

Printing History

edition number	printing number	update
1	1	New book.
1	2	Minor corrections.

Copyright © 2003 by Professional Publications, Inc. All rights reserved. No part of this publication may be reproduced, stored in a retrieval system, or transmitted, in any form or by any means, electronic, mechanical, photocopying, recording, or otherwise, without the prior written permission of the publisher.

Printed in the United States of America

Professional Publications, Inc.
1250 Fifth Avenue, Belmont, CA 94002
(650) 593-9119
www.ppi2pass.com

Library of Congress Cataloging-in-Publication Data
Schneiter, R. W.
 Six-minute solutions for civil PE exam environmental problems / R. Wane Schneiter.
  p. cm.
 ISBN 1-888577-89-4
 1. Environmental engineering--Examinations, questions, etc. I. Title.

TD157.S38 2003
628--dc22

2003066193

Table of Contents

ABOUT THE AUTHOR iv

PREFACE AND ACKNOWLEDGMENTS v

INTRODUCTION
    Exam Format vii
    This Book's Organization vii
    How to Use This Book vii

REFERENCES ix

BREADTH PROBLEMS
    Wastewater Treatment 1
    Aquatic Biology and Microbiology 2
    Solid and Hazardous Waste 3
    Groundwater and Well Fields 4

DEPTH PROBLEMS
    Wastewater Treatment 5
    Aquatic Biology and Microbiology 9
    Solid and Hazardous Waste 13
    Groundwater and Well Fields 17

BREADTH SOLUTIONS
    Wastewater Treatment 23
    Aquatic Biology and Microbiology 28
    Solid and Hazardous Waste 33
    Groundwater and Well Fields 38

DEPTH SOLUTIONS
    Wastewater Treatment 43
    Aquatic Biology and Microbiology 60
    Solid and Hazardous Waste 74
    Groundwater and Well Fields 95

About the Author

R. Wane Schneiter has BSCE and PhD degrees from Utah State University, is a Diplomate (Board Certified Environmental Engineer) in the American Academy of Environmental Engineers, and a PE in California and Virginia. He is currently the Benjamin H. Powell, Jr., Professor of Engineering in the Civil & Environmental Engineering Department at the Virginia Military Institute (VMI). Prior to joining the VMI faculty, Dr. Schneiter was a principal engineer with a consulting firm in the San Francisco Bay area and has been affiliated with engineering firms in California and Virginia. He advises industrial, governmental, and nongovernmental clients on a variety of diverse issues related to environmental and water resource engineering.

Preface and Acknowledgments

The Principles and Practice of Engineering examination (PE exam) for civil engineering, prepared by the National Council of Examiners for Engineering and Surveying (NCEES), is developed from sample problems submitted by educators and professional engineers representing consulting, government, and industry. PE exams are designed to test examinees' understanding of both conceptual and practical engineering concepts. Problems from past exams are not available from NCEES or any other source. However, NCEES does identify the general subject areas covered on the exam.

The topics covered in *Six-Minute Solutions for Civil PE Exam Environmental Problems* coincide with those subject areas identified by NCEES for the environmental engineering depth module of the civil PE exam. Included among these problem topics are wastewater treatment, biology (particularly as it pertains to aquatic environmental systems), solid and hazardous waste, and groundwater and well fields.

The problems presented in this book are representative of the type and difficulty of problems you will encounter on the PE exam. These are both conceptual and practical, and they are written to provide varying levels of difficulty. Though you probably won't encounter problems on the exam exactly like those presented here, reviewing these problems and solutions will increase your familiarity with the exam problems' form, content, and solution methods. This preparation will help you considerably during the exam.

Problems and solutions have been carefully prepared and reviewed to ensure that they are appropriate and understandable, and that they were solved correctly. If you find errors or discover an alternative, more efficient way to solve a problem, please bring it to PPI's attention so your suggestions can be incorporated into future editions. You can report errors and keep up with the changes made to this book, as well changes to the exam, by logging on to Professional Publications' website at www.ppi2pass.com and clicking on "Errata."

Thank you to the many persons in the editorial and production departments at PPI who contributed to the successful publication of this book. They are an enjoyable group to work with, are thorough and professional, and are dedicated to providing the best possible publication. Thanks are also due to James R. Sheetz, PE, DEE for this thorough technical review.

R. Wane Schneiter, PhD, PE, DEE

Introduction

EXAM FORMAT

The Principles and Practice of Engineering examination (PE exam) in civil engineering is an eight hour exam divided into a morning and an afternoon session. The morning session is known as the "breadth" exam, and the afternoon is known as the "depth" exam.

The morning session includes 40 problems from all of the five civil engineering subdisciplines (environmental, geotechnical, structural, transportation, and water resources), each subdiscipline representing about 20% of the problems. As the "breadth" designation implies, morning session problems are general in nature and wide-ranging in scope.

The afternoon session allows the examinee to select a "depth" exam module from one of the five subdisciplines. The 40 problems included in the afternoon session require more specialized knowledge than those in the morning session.

All problems from both the morning and afternoon sessions are multiple choice. They include a problem statement with all required defining information, followed by four logical choices. Only one of the four options is correct. Nearly every problem is completely independent of all others, so an incorrect choice on one problem typically will not carry over to subsequent problems.

Topics and the approximate distribution of problems on the afternoon session of the civil environmental exam are as follows.

Environmental: approximately 65% of exam problems

- Wastewater Treatment
- Aquatic Biology and Microbiology
- Solid and Hazardous Waste
- Groundwater and Well Fields

Geotechnical: approximately 10% of exam problems

- Subsurface Exploration and Sampling
- Engineering Properties of Soils
- Soil Mechanics Analysis

Water Resources: approximately 25% of exam problems

- Hydraulics
- Hydrology
- Water Treatment

For further information and tips on how to prepare for the civil environmental engineering PE exam, consult the *Civil Engineering Reference Manual* or Professional Publications' website, www.ppi2pass.com.

THIS BOOK'S ORGANIZATION

Six-Minute Solutions for Civil PE Exam Environmental Problems is organized into two sections. The first section, Breadth Problems, presents 20 environmental engineering problems of the type that would be expected in the morning part of the civil engineering PE exam. The second section, Depth Problems, presents 80 problems representative of the afternoon part of this exam. The two sections of the book are further subdivided into the topic areas covered by the environmental exam.

Most of the problems are quantitative, requiring calculations to arrive at a correct solution. A few are nonquantitative. Some problems will require a little more than 6 minutes to answer and others a little less. On average, you should expect to complete 80 problems in 480 minutes (8 hours), or spend 6 minutes per problem.

Six-Minute Solutions for Civil PE Exam Environmental Problems does not include problems related directly to geotechnical and water resources engineering, although problems from these subdisciplines will be included in the environmental exam. *Six-Minute Solutions for Civil PE Exam Geotechnical Problems* and *Six-Minute Solutions for Civil PE Exam Problems: Water Resources* provide problems for review in these areas of civil engineering.

HOW TO USE THIS BOOK

In *Six-Minute Solutions for Civil PE Exam Environmental Problems*, each problem statement, with its supporting information and answer choices, is presented in the same format as the problems encountered on the PE exam. The solutions are presented in a step-by-step sequence to help you follow the logical development of the correct solution and to provide examples of how you may want to approach your solutions as you take the PE exam.

Each problem includes a hint to provide direction in solving the problem. In addition to the correct solution, you will find an explanation of the faulty solutions leading to the three incorrect answer choices. The incorrect solutions are intended to represent common mistakes made when solving each type of problem. These may be simple mathematical errors, such as failing to square a term in an equation, or more serious errors, such as using the wrong equation.

To optimize your study time and obtain the maximum benefit from the practice problems, consider the following suggestions.

1. Complete an overall review of the problems and identify the subjects that you are least familiar with. Work a few of these problems to assess your general understanding of the subjects and to identify your strengths and weaknesses.

2. Locate and organize relevant resource materials. (See the references section of this book as a starting point.) As you work problems, some of these resources will emerge as more useful to you than others. These are what you will want to have on hand when taking the PE exam.

3. Work the problems in one subject area at a time, starting with the subject areas that you have the most difficulty with.

4. When possible, work problems without utilizing the hint. Always attempt your own solution before looking at the solutions provided in the book. Use the solutions to check your work or to provide guidance in finding solutions to the more difficult problems. Use the incorrect solutions to help identify pitfalls and to develop strategies to avoid them.

5. Use each subject area's solutions as a guide to understanding general problem-solving approaches. Although problems identical to those presented in *Six-Minute Solutions for Civil PE Exam Environmental Problems* will not be encountered on the PE exam, the approach to solving problems will be the same.

Solutions presented for each example problem may represent only one of several methods for obtaining a correct answer. Although we have tried to prepare problems with unique solutions, alternative problem-solving methods may produce a different, but nonetheless appropriate, answer.

References

The minimum recommended library for the civil exam consists of PPI's *Civil Engineering Reference Manual*. You may also find the following references helpful in completing some of the problems in *Six-Minute Solutions for Civil PE Exam Environmental Problems*.

Dean, J.A. *Lange's Handbook of Chemistry*. 15th ed. McGraw Hill, New York, N.Y. 1998.

Fetter, C.W. *Applied Hydrogeology*. 3rd ed. Macmillan College Publishing, New York, N.Y. 1994.

Fetter, C.W. *Contaminant Hydrogeology*. 2nd ed. Prentice Hall, Upper Saddle River, N.J. 1999.

Lide, D.R. *Handbook of Chemistry and Physics*. 82nd ed. CRC Press, Boca Raton, Fla. 2001.

Masters, G.M. *Introduction to Environmental Engineering and Science*. 2nd ed. Prentice Hall, Upper Saddle River, N.J. 1996.

McGhee, T.J. *Water Supply and Sewerage*. 6th ed. McGraw-Hill, New York, N.Y. 1991.

Metcalf & Eddy, Inc. *Wastewater Engineering: Treatment, Disposal, and Reuse*. 4th ed. McGraw-Hill, New York, N.Y. 2003.

Noll, K.E. *Fundamentals of Air Quality systems*. American Academy of Environmental Engineers, Annapolis, Md. 1999.

Ostler, N.K. (Ed.). *Introduction to Environmental Technology*. Prentice Hall, Englewood Cliffs, N.J. 1996.

Peavy, H.S., D.R. Rowe, and G. Tchobanoglous. *Environmental Engineering*. McGraw Hill, New York, N.Y. 1985.

Sawyer, C.N., P.L. McCarty, and G.F. Parkin. *Chemistry for Environmental Engineering*. 4th ed. McGraw-Hill, New York, N.Y. 1994.

Sincero, A.P. and G.A. Sincero. *Environmental Engineering: A Design Approach*. Prentice Hall, Upper Saddle River, N.J. 1996.

Breadth Problems

WASTEWATER TREATMENT

PROBLEM 1

What is the volatile suspended solids concentration of the following wastewater sample?

sample volume filtered (VF)	200 mL
sample volume evaporated (VD)	100 mL
mass of dried crucible and filter paper (MS)	25.439 g
mass of dry evaporation dish (MD)	275.41 g
mass of dried crucible, filter paper, and solids (MSS)	25.645 g
mass of dried evaporation dish and solids (MDS)	276.227 g
mass of ignited crucible, filter paper, and solids (MSI)	25.501 g
mass of ignited evaporation dish and solids (MDI)	276.201 g

(A) 260 mg/L
(B) 310 mg/L
(C) 720 mg/L
(D) 1000 mg/L

Hint: Find the solids that were ignited, not the ash that remains.

PROBLEM 2

What is the average hydraulic detention time for a rectangular tank with dimensions of 2.5 m by 15 m by 3.0 m deep receiving a flow of 900 m^3/d? The hydraulic efficiency of the tank is 83%.

(A) 2.3 h
(B) 2.5 h
(C) 3.0 h
(D) 3.6 h

Hint: The average detention time will be less than the theoretical detention time.

PROBLEM 3

Which of the following characteristics are included in the minimum national standards for secondary wastewater treatment under the Clean Water Act (CWA)?

   I.   suspended solids
   II. 5 day biochemical oxygen demand
   III. disinfection byproducts
   IV. dissolved solids

(A) I and II
(B) I and III
(C) II and IV
(D) III and IV

Hint: The CWA addresses discharges to receiving waters.

PROBLEM 4

What tank size is required to equalize the flow described by the following data?

time period (hr)	period average flow (10^6 gal/day)
0000–0400	1.39
0400–0800	3.21
0800–1200	4.05
1200–1600	2.63
1600–2000	3.91
2000–2400	1.98

(A) 4.8×10^5 gal
(B) 5.0×10^5 gal
(C) 2.1×10^6 gal
(D) 2.9×10^6 gal

Hint: Begin by preparing a table to construct a cumulative flow plot. The average flow is for each 4 hr period.

PROBLEM 5

Complete mix activated sludge has been selected for treatment of a wastewater.

flow rate	25×10^6 gal/day
reactor volume	5×10^6 gal
influent biochemical oxygen demand (BOD) concentration	224 mg/L
effluent BOD concentration	20 mg/L
reactor mixed liquor suspended solids concentration	3500 mg/L
recirculated solids concentration	12 000 mg/L
mean cell residence time	10 day
yield coefficient	0.5 g/g
endogenous decay rate constant	0.05/day

What is most nearly the recirculated solids flow rate required to maintain the mean cell residence time (MCRT)?

(A) 1.0×10^7 gal/day
(B) 1.8×10^7 gal/day
(C) 2.5×10^7 gal/day
(D) 6.0×10^7 gal/day

Hint: Begin by deriving an equation for the recirculated solids flow rate. Consider a mass balance around the clarifier.

PROBLEM 6

A wastewater treatment process wastes sludge at 50,000 gal/day. The wasted sludge contains 1.2% solids. What volume reduction can be realized by thickening and dewatering the sludge to 24% solids?

(A) 2500 gal/day
(B) 2600 gal/day
(C) 39,000 gal/day
(D) 48,000 gal/day

Hint: This problem can be solved by simple ratios and differences.

AQUATIC BIOLOGY AND MICROBIOLOGY

PROBLEM 7

Among the common types of pathogenic organisms listed, which of them is infrequently looked for in routine analysis, generally less susceptible to chlorination, and targeted for removal by filtration?

  I.   protozoa
  II. viruses
  III. bacteria
  IV. helminths

(A) I
(B) II and III
(C) II and IV
(D) I and IV

Hint: Generally, as microorganisms increase in size and complexity, especially in reproductive method, they are more likely to be resistant to chlorination and more likely to be removed by filtration.

PROBLEM 8

The coliform group is used as indicator organisms for pathogens because they have all of the following characteristics EXCEPT

(A) they apply to all types of water
(B) they are always present when pathogens are present, and normally absent otherwise
(C) they are easily detectable by routine analytical methods
(D) they are themselves pathogenic

Hint: Which characteristic would be undesirable?

PROBLEM 9

What is most nearly the 5 day biochemical oxygen demand (BOD_5) at 15°C of the water represented by the tabulated data? The samples were incubated for 5 days at 20°C. The rate coefficient is 0.17 d^{-1} (base 10).

bottle	sample volume (mL)	dissolved oxygen at $t = 0$ d (mg/L)	dissolved oxygen at $t = 5$ d (mg/L)
1	20	9.0	1.0
2	10	9.1	2.9
3	5	9.1	6.1
4	2	9.2	8.1

(A) 140 mg/L
(B) 150 mg/L
(C) 170 mg/L
(D) 180 mg/L

Hint: Review the data to see if the problem can be simplified by disregarding some of the bottles. Remember, you have to calculate ultimate BOD before applying the temperature correction.

PROBLEM 10

A wastewater treatment plant (WWTP) discharges raw sewage during periods of high rainfall. Typical discharge flows are 15×10^6 gal/day with dissolved oxygen concentrations of 1.20 mg/L. During these periods, the river flows at 2000 ft^3/sec with a dissolved oxygen concentration of 8.10 mg/L. What is most nearly the dissolved oxygen concentration in the river once complete mixing of the WWTP discharge has occurred?

(A) 4.65 mg/L
(B) 8.02 mg/L
(C) 8.05 mg/L
(D) 8.18 mg/L

Hint: Look at this as a mass balance problem.

PROBLEM 11

What is the lifetime risk to an adult from ingesting drinking water from a well contaminated with 100 ppb trichloroethylene (TCE) and 7.2 ppb 1,1-dichloroethylene (1,1-DCE)? The potency factor for TCE is 0.011 $(mg/kg \cdot d)^{-1}$ and for 1,1-DCE is 0.58 $(mg/kg \cdot d)^{-1}$.

(A) 32 in one million
(B) 64 in one million
(C) 76 in one million
(D) 770 in one million

Hint: Use EPA recommended exposure factors and be careful of units.

SOLID AND HAZARDOUS WASTE

PROBLEM 12

The per capita solid waste generation rate for the 175,000 residents of a city is 1.9 kg/d. The solid waste characteristics for the city are as follows.

component	mass (%)	component discarded moisture (%)	component discarded density (kg/m^3)
paper	31	6	85
garden	29	60	105
food	10	70	290
cardboard	9	5	50
wood	8	20	240
plastic	7	2	65
miscellaneous inert materials	6	8	480

What is the discarded (wet) density of the bulk waste?

(A) 91 kg/m^3
(B) 100 kg/m^3
(C) 140 kg/m^3
(D) 1300 kg/m^3

Hint: It is possible to calculate more than one density for the waste. Decide which density is being requested and note that it is being requested for the bulk waste.

PROBLEM 13

For the time study and route analysis presented as follows, how many residences can one truck and crew service in a single 8 hr day of curbside waste collection?

number of residences in community	4600
weekly as-discarded waste volume per residence	0.29 yd^3
average driving time between residences	14 sec
average pick-up/load time at each residence	32 sec
travel time from truck yard to route start	25 min
average travel time between route and transfer station	45 min
time to unload at transfer station	15 min
travel time from transfer station to truck yard	25 min
truck compacted waste capacity	8 yd^3
truck noncompacted equivalent capacity	20 yd^3

(A) 110 residences/day
(B) 210 residences/day
(C) 480 residences/day
(D) 630 residences/day

Hint: Organize the route by task and task time. Filling the truck is one task, so find the time to fill the truck once.

PROBLEM 14

A municipality with a population of 215,000 is under state mandate to recycle 25% of the solid waste generated by its citizens. The remaining 75% will be landfilled. The per capita waste generation rate is 4.6 lbm/day. The landfilled waste in-place maximum compacted density is 50 lbm/ft^3, and the soil-cover-to-compacted-waste ratio is 1:4.5 by volume. The landfill covers a rectangular area 1200 ft by 1600 ft. The maximum landfill height cannot exceed 80 feet with 1:1 side slopes. What is the operating life of the landfill?

(A) 5 yr
(B) 14 yr
(C) 21 yr
(D) 64 yr

Hint: Be careful applying the cover-to-fill ratio and distinguishing between the waste landfilled and the waste recycled.

PROBLEM 15

An electronics component manufacturer generates wastewater at 135 gal/min containing 12 mg/L of methylene chloride. A concentration of 100 μg/L was set as the treatment criteria for reuse of the water. The average water temperature is 25°C. The mass transfer coefficient using the selected packing material is 0.023.

What is most nearly the minimum required air flow rate if air stripping is the selected removal process? Assume a stripping factor of 3.5.

(A) 4.0 ft^3/min
(B) 63 ft^3/min
(C) 490 ft^3/min
(D) 900 ft^3/min

Hint: Units of Henry's constant are important to obtain the correct solution. Select these so the air-to-water ratio is unitless.

PROBLEM 16

The principal organic hazardous constituents (POHC) mass feed rate to an incinerator is 4.12 kg/h. The POHC mass rate from the incinerator to air pollution control equipment is 0.43 kg/h and from the stack is 0.00074 kg/h. What is the destruction and removal efficiency (DRE) for the POHC?

(A) 10.26%
(B) 89.56%
(C) 99.83%
(D) 99.98%

Hint: The correct solution requires that both destruction and removal be included.

GROUNDWATER AND WELL FIELDS

PROBLEM 17

What is the hydraulic conductivity of the aquifer for nonaqueous phase liquid (NAPL)?

hydraulic conductivity (water)	2.0×10^{-4} cm/s
aquifer temperature	$10°C$
density NAPL at $10°C$	0.92 g/cm^3
dynamic viscosity NAPL at $10°C$	0.066 g/cm·s

(A) 1.9×10^{-7} cm/s
(B) 3.6×10^{-5} cm/s
(C) 4.0×10^{-5} cm/s
(D) 1.8×10^{-4} cm/s

Hint: Distinguish between hydraulic conductivity and intrinsic permeability and review how they are related.

PROBLEM 18

An unconfined aquifer with a pumped well and an observation well is shown in the illustration.

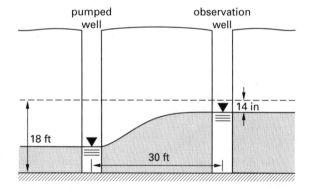

The pumped well diameter is 4 in, and the aquifer thickness is 18 ft with a hydraulic conductivity of 7.2 ft/day. The observation well is located 30 ft from the pumped well, and after 10 hr of pumping at 20 gal/min shows a steady-state drawdown of 14 in. What is the radius of influence of the pumped well?

(A) 2.3 ft
(B) 31 ft
(C) 38 ft
(D) 53 ft

Hint: Under steady-state conditions, the radius of influence is not a function of time.

PROBLEM 19

Bore-hole logs show interbedded soil layers with the following characteristics. What is the overall hydraulic conductivity through the soil layers?

layer	thickness (cm)	soil class
1	70	SP
2	109	GC
3	88	SM
4	46	SC

(A) 5.0×10^{-8} cm/s
(B) 1.3×10^{-7} cm/s
(C) 1.3×10^{-4} cm/s
(D) 5.0×10^{-4} cm/s

Hint: Only the soil classification is given. Try using typical values to estimate the hydraulic conductivity of each layer. Hydraulic conductivity and permeability were often used interchangeably.

PROBLEM 20

What is the approximate solute actual velocity for the site depicted in the illustration if the hydraulic conductivity is 0.83 ft/day, the soil porosity is 0.37, and the retardation factor is 1.94?

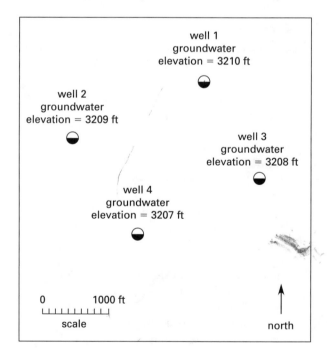

(A) 0.00056 ft/day
(B) 0.0015 ft/day
(C) 0.0021 ft/day
(D) 0.0029 ft/day

Hint: Be careful to distinguish between actual and Darcy velocity. Will the retardation factor increase or decrease the solute velocity compared to the groundwater velocity?

Depth Problems

WASTEWATER TREATMENT

PROBLEM 21

An industrial plant currently pays a surcharge for discharge of untreated wastewater to a city sewer system. Plans to construct a wastewater pretreatment system have included flow monitoring with the following results.

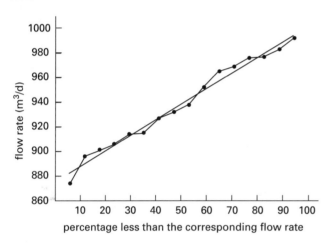

week	weekly average (m^3/d)
1	976
2	901
3	938
4	965
5	977
6	992
7	874
8	914
9	906
10	927
11	969
12	932
13	983
14	896
15	915
16	952

What flow rate should be selected for design so that the flow does not exceed the capacity of the pretreatment system more than 10% of the time?

(A) 850 m^3/d
(B) 890 m^3/d
(C) 990 m^3/d
(D) 1100 m^3/d

Hint: Are the results more useful in the tabular or graphic form?

PROBLEM 22

Wastewater treatment guidelines for a planned community require that wastewater treatment capacity be provided based on four population equivalents (PE) per home. The community will eventually include 1750 homes. What is the approximate biochemical oxygen demand (BOD) loading expected at the wastewater plant from the community?

(A) 88 lbm BOD/day
(B) 1400 lbm BOD/day
(C) 2200 lbm BOD/day
(D) 35,000 lbm BOD/day

Hint: Pay attention to units for all parameters.

PROBLEM 23

500 mL of wastewater with an initial pH of 7.3 is titrated with 0.03 N H_2SO_4. The 4.5 endpoint pH is reached when 14.5 mL of acid have been added. What is the concentration of the bicarbonate alkalinity?

(A) 22 mg/L as $CaCO_3$
(B) 29 mg/L as $CaCO_3$
(C) 44 mg/L as $CaCO_3$
(D) 72 mg/L as $CaCO_3$

Hint: The concentration of the acid used determines the amount of alkalinity neutralized and the initial pH defines what alkalinity species are dominant.

PROBLEM 24

Disinfection of a wastewater using aqueous chlorine at a pH of 8.5 and a temperature of 21°C requires 23 min to effect the desired percentage kill. How much time is required if the wastewater temperature is 17°C?

(A) 6.4 min
(B) 20 min
(C) 27 min
(D) 32 min

Hint: Use activation energy to correct for temperature.

PROBLEM 25

Which statement best defines priority pollutants as regulated under the Clean Water Act (CWA)?

(A) They are the most toxic of known chemicals.
(B) They are chemicals with relatively high toxicity and high production volume.
(C) They are chemicals that meet specific criteria of toxicity, flammability, corrosivity, or reactivity.
(D) They are chemicals associated with National Priorities List (NPL) sites.

Hint: Do not confuse provisions of the CWA with those of other regulations such as those associated with the Resource Conservation and Recovery Act (RCRA) and the Comprehensive Environmental Response, Compensation, and Liability Act (CERCLA).

PROBLEM 26

What is the required width of the rectangular horizontal-flow grit chamber where the following conditions apply?

flow rate	3.5×10^6 gal/day
depth	4 ft
mean particle diameter	0.22 mm
grit specific gravity	2.65
Camp constant	0.05
Darcy friction factor	0.03

(A) 0.60 ft
(B) 1.9 ft
(C) 8.8 ft
(D) 14 ft

Hint: What calculation or equation is suggested by the given information?

PROBLEM 27

What is most nearly the sludge volume index for a mixed liquor suspended solids (MLSS) suspension at an initial concentration of 2400 mg/L that settles to the 356 mL mark in a 1 L graduated cylinder after 30 min?

(A) 150 mL/g
(B) 270 mL/g
(C) 3700 mL/g
(D) 6700 mL/g

Hint: Sludge volume index units are uncommon.

PROBLEM 28

Bench test results for packing media evaluated to strip benzene from contaminated groundwater are summarized in the table. The tests are performed at 20°C. What is the mass transfer coefficient for the packing media and benzene at 8°C?

elapsed time (min)	benzene concentration in sample (μg/L)
0	978
2	569
4	303
6	153
8	76
10	36

(A) 0.19 min^{-1}
(B) 0.23 min^{-1}
(C) 0.30 min^{-1}
(D) 0.40 min^{-1}

Hint: What is the reaction order?

PROBLEM 29

Settling removal efficiency curves of a settling basin designed for a Type II suspension are presented in the illustration.

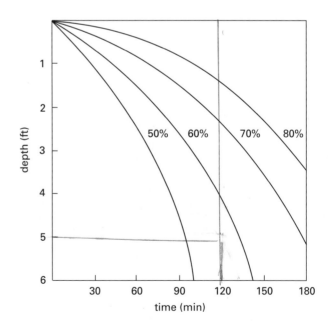

What is the removal efficiency at a depth of 5 ft and settling time of 120 min?

(A) 43%
(B) 57%
(C) 66%
(D) 71%

Hint: Incremental efficiency occurs at depth intervals.

PROBLEM 30

An anaerobic lagoon will pretreat a slaughterhouse wastewater that will be discharged to an existing facultative lagoon. Influent flow is 2.6×10^5 gal/day with a total biochemical oxygen demand (BOD) of 14 000 mg/L. The average waste temperature during winter months is 21°C and during summer months is 27°C. The winter loading rate is 12 lbm BOD/10^3 ft^3-day and the summer loading rate is 18 lbm BOD/10^3 ft^3-day. Site conditions limit the lagoon depth to 10 ft. What is the required total surface area of the anaerobic lagoon?

(A) 3.9 ac
(B) 4.6 ac
(C) 5.0 ac
(D) 5.8 ac

Hint: How do loading rates influence design?

PROBLEM 31

Reverse osmosis having the following characteristics was selected to treat a drinking water source with high total dissolved solids. How many pressure vessels are required to treat the water?

desired freshwater flow rate	16 000 m^3/d
permeate recovery	80%
salt rejection	92%
membrane flux rate	0.83 m^3/m^2·d
membrane packing density	800 m^2/m^3
membrane module volume	0.03 m^3
pressure vessel capacity	10 modules

(A) 63
(B) 80
(C) 87
(D) 100

Hint: Look for the membrane volume.

PROBLEM 32

A waste biological sludge is dewatered to 23% solids and then stabilized by raising the pH to 12.5 using lime dosed at 315 g Ca(OH)$_2$/kg dry solids. The plant wastes 18,000 gal of sludge daily at 9% solids. The locally available lime contains 22% inerts. What is most nearly the monthly mass of lime required to stabilize the sludge?

(A) 45 000 kg/month
(B) 58 000 kg/month
(C) 74 000 kg/month
(D) 190 000 kg/month

Hint: What information do the units used for lime dose reveal?

PROBLEM 33

An activated sludge plant operates with a mean cell residence time of 10 d to treat a flow of 18 925 m^3/d with an influent biochemical oxygen demand (BOD) concentration of 247 mg/L. The plant wastes sludge at 34 kg/d. What is most nearly the food-to-microorganism ratio for the plant?

(A) 0.014 d^{-1}
(B) 14 d^{-1}
(C) 140 d^{-1}
(D) 1400 d^{-1}

Hint: Begin with definitions.

PROBLEM 34

The following are selected characteristics of an activated sludge process bioreactor.

influent flow rate	27 000 m^3/d
influent biochemical oxygen demand (BOD)	281 mg/L
effluent BOD	20 mg/L
yield coefficient	0.53 g/g
endogenous decay rate constant	0.05 d^{-1}
mean cell residence time	8 d

What is the approximate daily mass of biomass produced in the bioreactor?

(A) 2700 kg/d
(B) 3700 kg/d
(C) 7000 kg/d
(D) 16 000 kg/d

Hint: Review the definition of the observed yield coefficient.

PROBLEM 35

What is most nearly the required media total surface area for a rotating biological contactor (RBC) process selected to treat the following wastewater?

flow rate	250,000 gal/day
influent total biochemical oxygen demand (BOD)	174 mg/L
effluent soluble BOD	30 mg/L

(A) 130,000 ft^2
(B) 150,000 ft^2
(C) 180,000 ft^2
(D) 360,000 ft^2

Hint: The solution to this problem is based on loading rates.

PROBLEM 36

Constructed wetlands for wastewater treatment (CWWT) are characterized as submerged flow (SF) and free water surface (FWS). An SF wetland would be preferred over an FWS wetland for all of the following conditions EXCEPT which one?

(A) An SF wetland provides improved odor and vector control over an FWS wetland.
(B) An SF wetland provides improved suspended solids removal over an FWS wetland.
(C) An SF wetland provides improved ammonia removal compared to an FWS wetland.
(D) An SF wetland is less susceptible to temperature and other climate extremes than are FWS wetlands.

Hint: Review the design and operating conditions of CWWT.

PROBLEM 37

A slaughterhouse wastewater at a flow rate of 5.0×10^5 gal/day contains solids that are best removed by flotation. The optimum air-to-solids ratio for the suspension concentration of 1200 mg/L is 1.3 ft^3/lbm. The wastewater temperature is 80°F and the atmospheric pressure is 1 atm. What is most nearly the air mass flow rate required by the flotation process?

(A) 480 lbm/day
(B) 8.9×10^4 lbm/day
(C) 4.0×10^5 lbm/day
(D) 3.3×10^8 lbm/day

Hint: Develop an equation using the air-to-solids ratio.

PROBLEM 38

Bench scale bioreactors operated to model activated sludge treatment of a wastewater produced the results presented in the table. The reactors are completely mixed without solids recycle.

reactor	influent biochemical oxygen demand (mg/L)	effluent biochemical oxygen demand (mg/L)	hydraulic residence time (d)	mixed liquor suspended solids (mg/L)
1	210	10	3.45	118
2	210	16	1.92	135
3	210	23	1.45	136
4	210	36	1.10	132

What is the value of the endogenous decay-rate coefficient?

(A) -1.2 d^{-1}
(B) -0.10 d^{-1}
(C) 0.14 d^{-1}
(D) 0.87 d^{-1}

Hint: Find an equation that includes the endogenous decay rate coefficient with the parameters given in the table.

PROBLEM 39

A municipality has selected an activated sludge process to provide a denitrified effluent. Selected data relevant to denitrification of the municipality's wastewater are

growth rate	0.38 d^{-1}
yield coefficient	0.81 g/g
wastewater temperature	16°C
methanol concentration	72 mg/L
influent nitrate concentration	29 mg/L
influent nitrite concentration	8 mg/L
half velocity constant for methanol	12 mg/L
half velocity constant for nitrogen	0.31 mg/L

What is the corrected maximum growth rate for denitrification?

(A) 0.011 d^{-1}
(B) 0.15 d^{-1}
(C) 0.17 d^{-1}
(D) 0.21 d^{-1}

Hint: Define growth rate, maximum growth rate, and corrected maximum growth rate.

PROBLEM 40

The following equations illustrate the difference in the amount of biomass produced by aerobic and anaerobic processes.

$$18 \text{ waste} + NH_4^+ HCO_3^- \rightarrow C_5H_7O_2N + 6.5 \text{ } CH_4 + 7.5 \text{ } CO_2 + 4 \text{ } H_2O$$

$$13 \text{ waste} + 3 \text{ } NH_4^+ + 10 \text{ } O_2 \rightarrow 3 \text{ } C_5H_7O_2N + 10 \text{ } HCO_3^- + CO_2 + 10 \text{ } H_2O$$

What is the ratio of biomass produced by the aerobic process to that produced by the anaerobic process?

(A) 1:3
(B) 3:1
(C) 4:1
(D) 8:1

Hint: It may look like insufficient information is given to reach a solution. Read the question carefully.

PROBLEM 41

What is most nearly the daily mass of oxygen required for aeration if nitrification is to occur for the following wastewater?

flow rate	5.0×10^6 gal/day
influent ammonia concentration	63 mg/L
effluent ammonia concentration	10 mg/L
influent five-day biochemical oxygen demand (BOD_5) concentration	356 mg/L
effluent BOD_5 concentration	30 mg/L
ratio of BOD_5 to ultimate BOD (BOD_u)	1:1.52
daily mass of biomass wasted	2800 lbm/day

(A) 17,000 lbm/day
(B) 20,000 lbm/day
(C) 27,000 lbm/day
(D) 31,000 lbm/day

Hint: Consider which conditions contribute to oxygen demand.

PROBLEM 42

Major provisions of the Clean Water Act (CWA) address all of the following EXCEPT

(A) categorical pretreatment standards for industrial effluents
(B) maximum contaminant levels for groundwater remediation
(C) national permit system for surface water discharges
(D) priority pollutants for regulation by discharge standards

Hint: The Clean Water Act (CWA) and the Safe Drinking Water Act (SDWA) are sometimes confused, but they are not the same things.

PROBLEM 43

An activated sludge process is characterized by the following parameters.

influent biochemical oxygen demand (BOD) from the primary clarifier	312 mg/L
influent ammonia nitrogen from the primary clarifier	51 mg/L as N
effluent total BOD	20 mg/L
effluent ammonia nitrogen	1 mg/L as N
maximum growth rate constant	0.50 d^{-1}
corrected maximum growth rate constant	0.41 d^{-1}
half velocity constant	2.6 mg/L
endogenous decay rate coefficient	0.07 d^{-1}

What is the minimum required mean cell residence time?

(A) 2.3 d
(B) 2.6 d
(C) 3.1 d
(D) 23 d

Hint: Consider how the mean cell residence time varies between nitrification and BOD removal.

AQUATIC BIOLOGY AND MICROBIOLOGY

PROBLEM 44

An industrial plant proposes to return noncontact cooling water to a river. The river water temperature upstream of the mixing zone is 6°C and the returned cooling water temperature is 20°C. The river flows at 280 m^3/s and the cooling water flow is 11 m^3/s. The reaeration and deoxygenation river constants are equal at 0.080 d^{-1}. What are the river constants downstream of the mixing zone?

(A) 0.075 d^{-1}
(B) 0.085 d^{-1}
(C) 0.19 d^{-1}
(D) 0.47 d^{-1}

Hint: Solve the problem in two parts.

PROBLEM 45

The relationship among assimilative capacity, stock pollutants, and fund pollutants can be best described by which of the following?

(A) Assimilative capacity, being associated with fund pollutants only, is not influenced by stock pollutants.
(B) Assimilative capacity, being associated with stock pollutants only, is not influenced by fund pollutants.
(C) Assimilative capacity is relatively low for fund pollutants and relatively high for stock pollutants.
(D) Assimilative capacity is relatively high for fund pollutants and relatively low for stock pollutants.

Hint: Think of examples of stock and fund pollutants.

PROBLEM 46

The dissolved oxygen and ultimate biochemical oxygen demand concentrations where a wastewater discharges to a stream are 9.3 mg/L and 9.8 mg/L, respectively. The stream flows at 0.3 ft/sec, and the stream water temperature is 8.6°C with reoxygenation and deoxygenation rate constants equal at 0.5 day^{-1} and 0.4 day^{-1}, respectively. Approximately how far downstream from the discharge point should the monitoring station be located to detect the maximum oxygen deficit caused by the discharge?

(A) 0.079 mi
(B) 2.4 mi
(C) 7.9 mi
(D) 180 mi

Hint: This is a critical-time problem.

PROBLEM 47

A microbial system is known to follow the Monod kinetic model. For this system, the specific growth rate is 6 d^{-1} and the substrate concentration is 16 mg/L. What is the maximum specific growth rate?

(A) 3 d^{-1}
(B) 4 d^{-1}
(C) 9 d^{-1}
(D) 12 d^{-1}

Hint: Consider the special case when the half-velocity coefficient is defined.

PROBLEM 48

Arsenic, cadmium, and fluoride have been detected in the soil of a city park at 1.3 ppb, 0.96 ppb, and 0.42 ppb, respectively. The oral route reference dose (RfD) for arsenic is 0.0003 mg/kg·d, for cadmium is 0.0005 mg/kg·d, and for fluoride is 0.0003 mg/kg·d. What is the hazard index if the exposed population is children who may ingest the soil?

(A) 1.1×10^{-12}
(B) 8.3×10^{-6}
(C) 0.000 10
(D) 0.51

Hint: The hazard index is for noncarcinogenic exposure.

PROBLEM 49

Biochemical oxygen demand (BOD) and theoretical oxygen demand (ThOD) for selected chemical wastes are presented in the following table. Which chemical waste listed in the table is the most likely to be biologically degradable?

	BOD (g/g)	ThOD (g/g)
chemical waste A	2.15	2.52
chemical waste B	1.34	2.15
chemical waste C	1.85	1.92
chemical waste D	1.64	2.91

(A) chemical waste A, because it has the greatest BOD
(B) chemical waste B, because it has the least BOD
(C) chemical waste C, because it has the greatest BOD to ThOD ratio
(D) chemical waste D, because it has the smallest BOD to ThOD ratio

Hint: Review the definitions of BOD and ThOD.

PROBLEM 50

What is the 96-h LC50 for the chemical represented by the toxicity test results summarized in the table? Twenty fathead minnow are used for each dilution. The predilution chemical concentration was 1.63 mg/L.

dilution (%)	survivors at 96-h
98	19
96	12
92	7
84	0

(A) 0.088 mg/L
(B) 0.10 mg/L
(C) 0.30 mg/L
(D) 1.5 mg/L

Hint: Plot the results.

PROBLEM 51

Which one of the following would likely NOT be effective for controlling algae in wastewater effluents?

(A) aeration
(B) microscreening
(C) nitrification/denitrification
(D) chlorination

Hint: Consider the treatment function of each choice.

PROBLEM 52

What is the corrected maximum growth rate for the nitrifying system described by the following characteristics?

pH	6.4
temperature	17°C
dissolved oxygen concentration	7.2 mg/L
yield coefficient	0.23 g/g
endogenous decay rate constant	0.05 d^{-1} at 17°C
growth rate constant	2.1 d^{-1} at 17°C

(A) 0.046 d^{-1}
(B) 0.17 d^{-1}
(C) 0.48 d^{-1}
(D) 0.72 d^{-1}

Hint: One or more typical parameter values are assumed.

PROBLEM 53

Surface water runoff with an average ammonia nitrogen concentration of 1.2 mg/L contributes 735,000 ft^3 per year to a lake with a constant volume of 1.9×10^6 ft^3. The ammonia nitrogen concentration is 0.26 mg/L in the stream flowing out of the lake. What is the oxygen demand exerted on the lake from the ammonia nitrogen?

(A) 25 kg O$_2$/yr
(B) 46 kg O$_2$/yr
(C) 91 kg O$_2$/yr
(D) 230 kg O$_2$/yr

Hint: Apply mass-balance principles and the nitrogenous oxygen demand.

PROBLEM 54

Results of a multiple-tube fermentation test are presented in the table. What is the best approximation of the most probable number (MPN)?

volume (mL)	positives	negatives
10	5	0
1	3	2
0.1	2	3
0.01	1	4

(A) 2/100 mL
(B) 100/100 mL
(C) 140/100 mL
(D) 170/100 mL

Hint: Use the simplest method to approximate the most probable number (MPN).

PROBLEM 55

A stream segment below a stormwater discharge exhibits the following characteristics.

saturated dissolved oxygen concentration	10.9 mg/L
mixed ultimate biochemical oxygen demand (BOD$_u$) at the discharge	7.2 mg/L
dissolved oxygen deficit at the discharge point	3.2 mg/L
reaeration constant	0.07 d^{-1}
deoxygenation constant	0.04 d^{-1}

Which oxygen sag curve represents the stream's dissolved oxygen profile below the discharge?

(A)

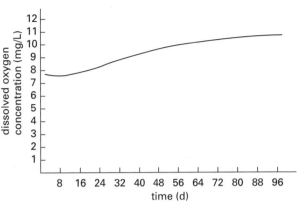

(B)

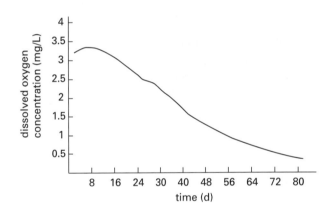

(C)

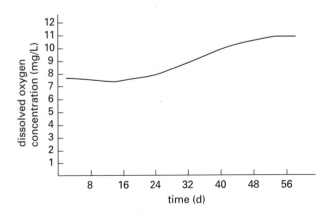

(D)

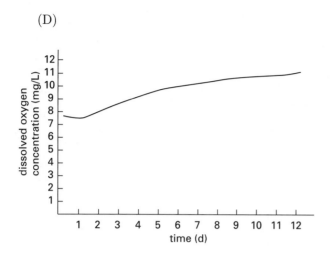

Hint: Oxygen sag curves are characterized by a typical profile.

PROBLEM 56

Characteristics of a wastewater treatment process include the following.

influent five-day biochemical oxygen demand (BOD_5)	217 mg/L
influent ultimate biochemical oxygen demand (BOD_u)	312 mg/L
effluent BOD_5	20 mg/L
effluent total suspended solids	18 mg/L
biodegradable fraction of effluent total suspended solids	0.62

What is the concentration of the influent soluble BOD_5 that escapes treatment?

(A) 4.2 mg/L
(B) 9.0 mg/L
(C) 12 mg/L
(D) 20 mg/L

Hint: Determine the relationship between effluent solids and effluent biochemical oxygen demand.

PROBLEM 57

Effluent limits for a proposed wastewater discharge to a river allow 65% recovery of the river's dissolved oxygen (DO) concentration within three days. The average river flow is 37 ft³/sec with an ultimate biochemical oxygen demand (BOD_u) of 6 mg/L upstream of the discharge. The illustration shows the dissolved oxygen profile for the river as a function of mixed-flow BOD_u (L_o). The projected discharge flow is 12×10^6 gal/day (12 MGD). What is the maximum allowable BOD_u of the discharge?

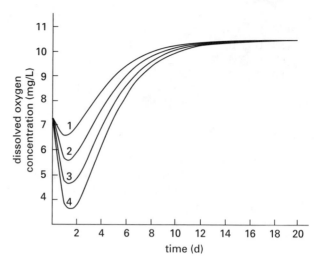

1 L_o = 10 mg/L
2 L_o = 14 mg/L
3 L_o = 18 mg/L
4 L_o = 22 mg/L

(A) 14 mg/L
(B) 30 mg/L
(C) 39 mg/L
(D) 54 mg/L

Hint: Start by examining the illustration. What information does it provide?

PROBLEM 58

A chemical formula typically used to represent biomass is given as $C_5H_7O_2N$. What is most nearly the daily nitrogen requirement if the wastewater contains 310 mg/L acetic acid and the flow rate is 12 000 m³/d?

(A) 140 kg/d
(B) 170 kg/d
(C) 230 kg/d
(D) 430 kg/d

Hint: Begin with the mole ratio.

PROBLEM 59

A constructed wetland for wastewater treatment (CWWT) is proposed for a community of 800 people. The CWWT design criteria imposed by the local health department are 6.0 m²/PE and 0.2 m³/PE·d with a minimum bed depth of 0.75 m. What is most nearly the empty-bed hydraulic residence time for the CWWT?

(A) 0.028 d
(B) 0.044 d
(C) 23 d
(D) 40 d

Hint: How is population equivalent (PE) used in the design criteria?

PROBLEM 60

Construction equipment has punctured a pipeline that runs parallel to a 6 ft wide concrete-lined drainage ditch.

The puncture is 1.8 miles upstream of where the ditch discharges into a saltwater marsh. 200 gal of a chemical were released from the puncture. The ditch normally flows at 3 ft/sec with a water depth of 2 ft. The average slope of the ditch channel is 1%. The specific gravity of the chemical is 0.9 and the ditch dispersion coefficient is 0.3 day^{-1}. What will be the maximum concentration of the chemical when it reaches the marsh?

(A) 6000 mg/L
(B) 6700 mg/L
(C) 6800 mg/L
(D) 8500 mg/L

Hint: Look for references regarding an instantaneous release into a river, channel, or estuary.

PROBLEM 61

A wastewater treatment plant experienced an emergency bypass of 4.2 million gallons of untreated domestic wastewater into a freshwater stream. How much time is required for 90% of the bacteria in the initial release to die.

(A) 0.11 d
(B) 0.69 d
(C) 2.3 d
(D) 10 d

Hint: Determine the reaction order and find an appropriate rate coefficient for bacterial die-off.

SOLID AND HAZARDOUS WASTE

PROBLEM 62

A solid waste is characterized as follows.

flash point	135°F
pH	8.1
lead concentration	4.3 mg/L by toxicity characteristic leaching procedure (TCLP)
toxaphene concentration	5.8 μg/L by TCLP
reactivity	nonreactive

Does the waste meet regulatory criteria for classification as a hazardous waste?

(A) Yes, the toxaphene concentration classifies the waste as hazardous waste on the basis of toxicity.
(B) Yes, the waste satisfies the ignitability criteria for classification as hazardous waste.
(C) No, the waste satisfies some, but not all, of the regulatory criteria for hazardous waste classification.
(D) No, the waste does not satisfy any of the regulatory criteria for hazardous waste classification.

Hint: See Title 40 of the Code of Federal Regulations, Secs. 261.20 to 261.24.

PROBLEM 63

The solid waste characteristics for a city are shown in the table.

component	discarded mass (%)	discarded moisture (%)	discarded density (lbm/ft^3)
paper	38	6	5
garden	25	60	7
food	9	70	18
cardboard	8	5	3
wood	8	20	15
plastic	7	2	4
inert materials	5	8	30

The per capita solid waste generation rate for the 100,000 residents of the city is 5 lbm/day.

The city wants to investigate potential revenue sources from selling composted mulch composed of all of the garden and food waste and some of the paper waste. Paper should contribute 10% (by volume) of the finished mulch. The mulch would be sold at 40% moisture for $10/yd^3.

What is the annual revenue potential from the sale of the mulch?

(A) $2.0 million/yr
(B) $2.6 million/yr
(C) $2.9 million/yr
(D) $3.1 million/yr

Hint: Consider each mulch component separately to determine the total mulch volume.

PROBLEM 64

A city desires to implement a curbside recycling program for newspapers, plastic containers, aluminum cans, steel cans, and glass. Residents will be required to segregate the waste, bundle newspapers, and crush aluminum and steel cans and plastic containers. No compaction will occur on the truck and the waste will remain segregated. The total (recyclable and non-recyclable) solid waste generation rate is 1.3 kg/person·d. On average, there are 3.6 residents per stop and collection would occur once weekly.

What truck capacity is required for the city to be able to complete collection on one route of 100 stops without having to unload until the end of the day?

component	average of total discarded (%)	average curbside density (kg/m^3)
newspaper	16	200
aluminum	1.3	240
steel	1.9	405
plastic	5.7	265
glass	4.1	180

(A) 1 m³
(B) 2 m³
(C) 4 m³
(D) 5 m³

Hint: How is the truck volume related to the weekly volume?

PROBLEM 65

A city collects solid waste from 435 commercial dumpsters. The dumpsters have an uncompacted capacity of 3 yd³ and, when filled, contain about 1000 lbm of mixed waste. The trucks that collect from the dumpsters have a capacity of 8 yd³ of waste compacted to 1200 lbm/yd³. Collection occurs once weekly.

It requires 2 min to empty each dumpster, and travel time between dumpsters requires an average of 6 min. Filled trucks dump at a centrally located transfer station, requiring a round trip including dumping time of 38 min from the end of one route to the beginning of the next. The trucks are parked at the transfer station at the end of each 8 hr work day.

How many dumpsters can one truck empty in a single day?

(A) 18 dumpsters/day
(B) 21 dumpsters/day
(C) 36 dumpsters/day
(D) 50 dumpsters/day

Hint: Find the number of dumpsters in a single compacted truckload.

PROBLEM 66

A city generates 72 tons of municipal solid waste per day that is to be incinerated. The waste has a heating value of 11 000 kJ/kg, is 12% ash, and has total combined moisture and hydrogen water of 48%. Heat loss to ash is 400 kJ/kg and radiation losses are expected to be 0.0035 kJ/kg. What is the net heat produced from incinerating the waste?

(A) 2.9×10^8 kJ/d
(B) 6.4×10^8 kJ/d
(C) 7.1×10^8 kJ/d
(D) 1.4×10^9 kJ/d

Hint: Calculate each type of heat loss individually.

PROBLEM 67

A city of 230,000 people generates solid waste at an average per capita rate of 2.5 lbm/day. The waste is landfilled in bales 3 ft by 3 ft by 6 ft that are stacked in cell layers three bales high. The bales each weigh 3500 lbm. The capacity of a single baling machine is 16 bales per hour. How many baling machines are required if the city wishes to limit the landfill baling operation to 40 hr per week?

(A) 1 machine
(B) 2 machines
(C) 7 machines
(D) 9 machines

Hint: Pay close attention to time.

PROBLEM 68

An industrial wastewater discharge of 2.5×10^5 gal/day contains n-butylphthalate at an average concentration of 148 mg/L. The required removal efficiency is 99%. The isotherm constants for n-butylphthalate are an intercept of 220 mg/g and slope of 0.45. Which of the following options is the smallest granular activated carbon (GAC) adsorber that can provide a minimum carbon change-out period of 14 days?

(A) 2000 lbm
(B) 4000 lbm
(C) 10,000 lbm
(D) 20,000 lbm

Hint: What do the intercept and slope units suggest for an isotherm equation?

PROBLEM 69

A plating shop produces 164 m³/d of wastewater containing cadmium at 31 mg/L as Cd^{2+}, zinc at 13 mg/L as Zn^{2+}, nickel at 21 mg/L as Ni^{2+}, and chromium at 130 mg/L as CrO_4^{2-}. Ion exchanger characteristics selected to treat the wastewater are shown in the following table.

characteristic	cation exchanger	anion exchanger
regenerant	H_2SO_4	NaOH
dosage (kg/m³)	192	76
concentration (%)	5	10
regenerant density (g/cm³)	1.44	1.15
hydraulic loading rate (m³/m³·min)	0.020	0.020
resin capacity (equiv/L)	1.5	3.7

What is most nearly the required regeneration period for the exchanger used to recover the cations?

(A) 7 min
(B) 11 min
(C) 33 min
(D) 130 min

Hint: Are cations exchanged with cationic or anionic resins?

DEPTH PROBLEMS 15

PROBLEM 70

A hazardous waste generated at 50,000 gal/day and with a pH of 1.6 is treated by simple neutralization using 0.005 N sodium hydroxide. What is the daily sodium hydroxide feed rate?

(A) 0.015 mL/d
(B) 3.8×10^3 mL/d
(C) 1.5×10^5 mL/d
(D) 9.5×10^{12} mL/d

Hint: Consider the relationship between pH and pOH.

PROBLEM 71

The loading rate for a municipal solid-waste incinerator is 650 000 kJ/m²·min. The municipality generates 325 000 kg of waste daily with a bulk discarded energy content of 890 000 kJ/m³ and bulk discarded density of 98 kg/m³. What is the required minimum bed area for the incinerator?

(A) 1.4 m²
(B) 3.2 m²
(C) 76 m²
(D) 30 000 m²

Hint: Be careful with units.

PROBLEM 72

Leachate is expected to accumulate on the upper surface of a landfill that is constructed with a natural clay stratum as the liner. The clay stratum has a diffusion coefficient of 8.7×10^{-9} m²/s and a tortuosity of 0.6. The chloride concentration in the leachate against the upper surface of the clay is expected to reach 12 000 mg/L. If chloride is used as the tracer, how thick should the clay stratum be to prevent the chloride concentration from reaching 100 mg/L on the underside of the liner until 100 yr have past?

(A) 8 m
(B) 15 m
(C) 20 m
(D) 61 m

Hint: The solution uses the effective diffusion.

PROBLEM 73

Contaminated site remediation options may include a "no action," or "no direct cleanup," alternative. Which of the following statements are most applicable to the no action alternative?

I. The responsible party is likely released from any future responsibility for the site.
II. Ongoing site monitoring may be required to ensure that conditions remain static.
III. Deed or future use restrictions may be imposed on the property.
IV. Remediation is exceptionally difficult or excessive costs may be incurred.

(A) I and IV
(B) I, II, and III
(C) II and III
(D) II, III, and IV

Hint: What kinds of sites may be candidates for no action?

PROBLEM 74

A municipality of 75 000 people generates solid waste at 1.9 kg/person·d with the proportion of individual waste components listed in the table.

component	dry mass (%)	density (kg/m³)
paper and paper products	36	140
yard waste	18	120
food waste	9.0	300
ferrous metals	5.1	160
non-ferrous metals	4.3	240
plastics	7.6	130
glass	6.9	350
wood	3.9	220
textiles	2.1	60
rubber	3.2	130
miscellaneous inert materials	3.9	480

Regional markets make recycling potentially feasible for paper and paper products, all metals, glass, and plastics. Because local regulations allow city authorities to assess fines for failure to separate and process (i.e., clean, crush, bundle) recyclable wastes for curbside collection, participation in the recycling program is expected to be 92%. What percentage of the total waste volume can potentially be recycled?

(A) 41%
(B) 54%
(C) 59%
(D) 64%

Hint: Since the percent of the total volume is desired, does a quick solution exist?

PROBLEM 75

A city anticipates the need for a solid waste transfer station within the coming few years. The amortized capital and operating cost of the transfer station is expected to be about $3.87 per ton of waste, and will include baling the waste. The city currently spends $0.061/ton-min for direct-haul to the landfill, but this cost will continue to increase as the city population grows and new development occurs. The city will spend $0.016/ton-min for hauling the baled waste from the transfer station to the existing landfill. The average travel time between the landfill and the transfer station or collection route is 72 min. At what direct-haul cost will it be more economical to build and operate the transfer station than to continue with direct-haul?

(A) $0.038/ton-min
(B) $0.054/ton-min
(C) $0.070/ton-min
(D) $0.11/ton-min

Hint: Find the breakeven point.

PROBLEM 76

Two alternative sites are proposed for a municipal solid waste landfill. The sites have been rated against criteria on a weighted scale from 1 (unimportant/poor) to 4 (very important/excellent) as shown in the table. What is the weighted average for the most desirable site?

category	criteria	weighting factor	site 1 rating	site 2 rating
location	haul distance	2	2	3
	access routes	4	3	1
	land value	3	2	2
soil/geology	permeability	3	3	4
	heterogeneities	4	2	3
	cover quantities	3	3	2
	seismic activity	4	4	4
groundwater	quality	2	3	3
	gradient	1	2	3
	depth	2	4	3
hydrology	drainage pattern	3	3	2
	streams	2	3	2
community	population	3	2	3
	land uses	4	4	3
	opposition	4	3	2

(A) 0.98
(B) 2.6
(C) 2.9
(D) 8.6

Hint: Review weighted average analysis.

PROBLEM 77

An abandoned industrial site with soil and groundwater contaminated by an organic solvent has the following characteristics.

gradient	0.00063
effective porosity	0.38
intrinsic permeability	1.1×10^{-5} mm^2
soil total organic carbon	485 mg/kg
soil bulk density	1.8 g/cm^3
soil-water partition coefficient	173 mL/g
temperature	10°C

What is the velocity of the dissolved organic solvent?

(A) 2.2×10^{-11} m/d
(B) 5.5×10^{-5} m/d
(C) 0.0084 m/d
(D) 0.012 m/d

Hint: Be careful to use correct definitions for the terms used in the equations.

PROBLEM 78

A plating line uses ion exchange to recover chromic acid, as CrO_3, from a 15% solution generated at 1000 m^3/d. The exchange resin deteriorates rapidly at CrO_3 concentrations greater than 11%. The recovered solution generated at 350 m^3/d contains 42% CrO_3, and to meet reuse specifications, the recovered CrO_3 must be diluted to a 3.5 molar concentration. What daily volume of dilution water is required by the ion exchange system?

(A) 430 m^3/d
(B) 1300 m^3/d
(C) 1800 m^3/d
(D) 2200 m^3/d

Hint: Find the dilution ratio for each step.

PROBLEM 79

A waste discharged at 50,000 gal/day contains Cr(VI) at 165 mg/L. Sodium metabisulfate, sulfuric acid, and caustic soda are used as reagents to produce the following reactions.

$$4CrO_3 + 3Na_2S_2O_5 + 3H_2SO_4$$
$$\rightarrow 3Na_2SO_4 + 2Cr_2(SO_4)_3 + 3H_2O$$

$$Cr_2(SO_4)_3 + 6NaOH$$
$$\rightarrow 2Cr(OH)_3 + 3Na_2SO_4$$

What is the daily dry mass of chromium sludge produced?

(A) 32 kg/d
(B) 65 kg/d
(C) 120 kg/d
(D) 31 000 kg/d

Hint: Does the solution require mole ratios or mass ratios?

PROBLEM 80

A rural county has roll-off boxes for solid waste collection at 34 locations. The average location includes three boxes each with an 18 yd^3 capacity. The boxes are emptied twice monthly. Average driving time between any two locations is 27 min and the average travel time from any location to the landfill is 45 min.

The county is considering using smaller, 6 yd^3 dumpsters that can be emptied into a compaction truck. The compaction truck capacity is 12 yd^3 with a compaction factor of 3.0. For the same collection schedule, approximately how much driving time will be saved if compaction trucks and dumpsters are used instead of the roll-off boxes?

(A) 36 hr/mo
(B) 61 hr/mo
(C) 69 hr/mo
(D) 140 hr/mo

Hint: How many dumpsters are equivalent to a roll-off box?

PROBLEM 81

A wastewater treatment plant plans to offset energy costs by augmenting its natural gas use with methane gas recovered from its anaerobic digesters. A mixture of 70% methane and 30% natural gas would be used. The plant currently pays $0.020/ft^3 for natural gas and expects an operating and amortized capital cost of $0.0080/ft^3 of digester gas to scrub the methane before blending. Total gas production by the digesters is expected to peak at 38,000 ft^3/day, 65% of which would be methane. What will be the approximate annual savings over the current natural gas costs if all the available methane gas, when blended with the natural gas, precisely meets the needs of the plant?

(A) $69,000/yr
(B) $110,000/yr
(C) $170,000/yr
(D) $310,000/yr

Hint: Be careful with methane and blended gas fractions. Work the problem in parts.

GROUNDWATER AND WELL FIELDS

PROBLEM 82

The results of an ion analysis of a groundwater sample are summarized in the table. Are all ions that are likely to be present at significant concentration included in the analysis?

ion	concentration (mg/L)
Ca^{2+}	128
Mg^{2+}	66
SO_4^{2-}	83
Cl^-	21
NO_3^-	14
HCO_3^-	279
Na^+	7

(A) Yes, anions and cations balance.
(B) Yes, analysis is deficient in cations.
(C) No, analysis is deficient in anions.
(D) No, analysis is deficient in cations.

Hint: The units used to express concentration are important to determine the adequacy of the analysis.

PROBLEM 83

The intrinsic permeability of a soil is 1.4×10^{-6} in^2. The groundwater gradient is 0.00035 and the soil effective porosity is 0.42. What is the rate of advection if the groundwater temperature is 45°F?

(A) 0.62 ft/day
(B) 1.5 ft/day
(C) 15 ft/day
(D) 210 ft/day

Hint: The intrinsic permeability is a property of the soil only.

PROBLEM 84

The results of a soil adsorption isotherm test using groundwater contaminated with an organic chemical are presented in the following illustration.

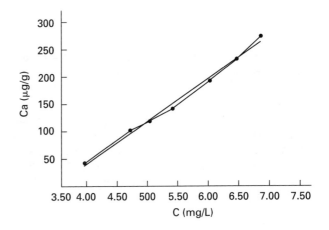

The soil effective porosity is 0.43, the soil bulk density is 1.68 g/cm^3, and the soil total organic carbon (TOC) is 271 mg/kg. What is the approximate relative velocity of the organic chemical to the groundwater?

(A) 9.0×10^{-7} v_{gw}
(B) 3.3×10^{-6} v_{gw}
(C) 0.0033 v_{gw}
(D) 0.95 v_{gw}

Hint: What information does the isotherm plot provide?

PROBLEM 85

A chemical mixture released to the groundwater contains 79% tetrachloroethene (PCE), 11% trichloroethene (TCE), and 10% trans-1,2-dichlorethene (t-1, 2-DCE). What is most nearly the equilibrium concentration of PCE in solution with the groundwater?

(A) 72 mg/L
(B) 110 mg/L
(C) 120 mg/L
(D) 150 mg/L

Hint: How is the solution influenced by the water solubility of the chemicals?

PROBLEM 86

What is the longitudinal hydrodynamic dispersion of an aquifer with the following characteristics?

gradient	0.0012 m/m at S 30° W
hydraulic conductivity	0.42 m/d
effective porosity	0.36
dynamic dispersivity	872 m

(A) 0.17 m²/d
(B) 0.44 m²/d
(C) 0.80 m²/d
(D) 1.2 m²/d

Hint: Be careful to distinguish dispersion from diffusion and Darcy velocity from actual velocity.

PROBLEM 87

A small pest-control business routinely discharges pesticide-contaminated water to a drainage ditch when washing their equipment. The ditch infiltrates to a shallow aquifer with a bulk groundwater velocity of 172 cm/d. The concentration of the pesticide in the aquifer just below the ditch is 0.182 mg/L. How long will be required for the pesticide to reach a drinking water well located in a direct line 1600 m downgradient of the source at its maximum contaminant level (MCL) of 1.0 μg/L?

(A) 0.8 d
(B) 84 d
(C) 860 d
(D) 5300 d

Hint: Make simplifying assumptions and choose an equation for the appropriate boundary conditions.

PROBLEM 88

A saturated soil profile is characterized in the table. The groundwater elevation difference between two wells, one that screens layer 1 and the other that screens layer 4, is 14 cm. What is the vertical groundwater flow between the screened layers?

layer	thickness (cm)	hydraulic conductivity (cm/s)
1	130	0.0090
2	180	0.017
3	270	0.036
4	65	0.011
5	110	0.020

(A) 2.18×10^{-5} m³/s for 1 m² of aquifer area
(B) 2.35×10^{-5} m³/s for 1 m² of aquifer area
(C) 2.41×10^{-5} m³/s for 1 m² of aquifer area
(D) 2.56×10^{-5} m³/s for 1 m² of aquifer area

Hint: How is the overall hydraulic conductivity calculated?

PROBLEM 89

Which of the following are most likely NOT potentially suitable methods for in situ restoration of an aquifer contaminated with inorganic chemicals?

I. chemical oxidation using hydrogen peroxide as the oxidizing agent
II. biodegradation enhanced by nutrient augmentation and air sparging
III. chemical oxidation through superchlorination
IV. physical containment using a reactive or adsorption media placed as cut-off walls

(A) I
(B) I and III
(C) II and III
(D) IV

Hint: Consider impacts from possible byproduct formation.

PROBLEM 90

The biological degradation of an organic chemical in groundwater is represented by the following equation.

$$26CH_3COO^- + 6NH_4^+ + 21O_2$$
$$\rightarrow 6C_5H_7O_2N + 18H_2O + 2CO_2 + 20HCO_3^-$$

The groundwater contains ammonia from septic tank leachate at 9.7 mg/L and the contaminant is present at 818 mg/L. If no nitrogen augmentation is included, by approximately how much will the contaminant concentration be reduced through biodegradation?

(A) 7.4 mg/L
(B) 58 mg/L
(C) 140 mg/L
(D) 810 mg/L

Hint: Start with ammonia.

PROBLEM 91

Flow lines for a contaminated groundwater extraction system are shown in the illustration. What circled letters on the illustration identify stagnation zones in the extraction system?

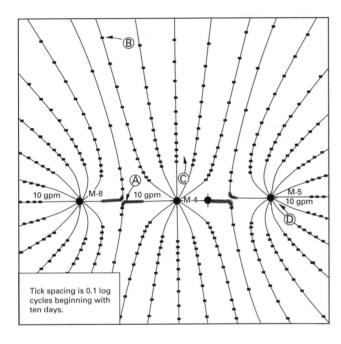

(A) A
(B) B
(C) C
(D) D

Hint: The flow lines and tick marks tell the story.

PROBLEM 92

A coffee processing plant uses methylene chloride for preparing decaffeinated product. The wastewater from the process contains methylene chloride at a concentration of 1.62 mg/L in a waste stream of 1000 m³/d. What is the packing height of the air stripping column required to recover 99.6% of the methylene chloride from the wastewater? Design parameters for the stripping tower and characteristics for methylene chloride follow.

Henry's law constant   0.078
mass-transfer coefficient 0.013 s^{-1}
stripping factor   3
hydraulic loading rate $0.022 \text{ m}^3/\text{m}^2 \cdot \text{s}$

(A) 4.5 m
(B) 13 m
(C) 14 m
(D) 16 m

Hint: Find the number and height of transfer units.

PROBLEM 93

The illustration represents a chromatograph for gasoline sampled from an underground storage tank.

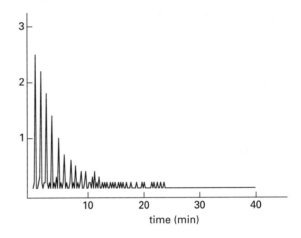

Assume the underground storage tank leaked some of the gasoline to groundwater. Which of the following chromatographs would most likely represent the same gasoline as sampled from a monitoring well?

(A)

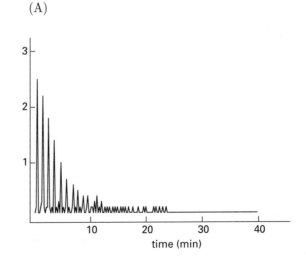

(B)

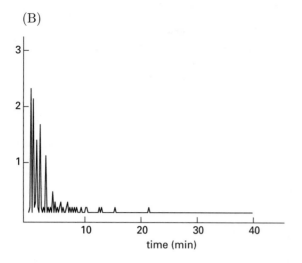

(C)

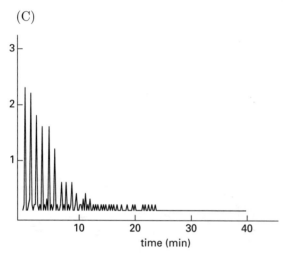

(D)

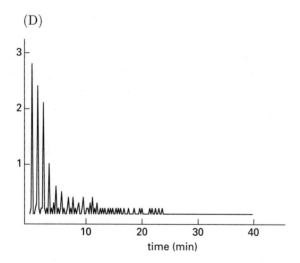

Hint: How would weathering influence the chromatograph?

PROBLEM 94

In a soil-groundwater system, hydrodynamic dispersion is a combination of mechanical dispersion, commonly referred to simply as dispersion, and of diffusion. In a low permeability soil with very shallow groundwater gradient, which component of hydrodynamic dispersion will dominate and what will be the general shape of a solute plume?

(A) Dispersion dominates. The plume has a pronounced elongated shape.
(B) Dispersion dominates. The plume has a relatively circular shape.
(C) Diffusion dominates. The plume has a pronounced elongated shape.
(D) Diffusion dominates. The plume has a relatively circular shape.

Hint: Define dispersion and diffusion.

PROBLEM 95

Which of the following conditions contribute to an increase in the residual saturation of a nonaqueous phase liquid (NAPL) released to the soil?

I. pre-wetting of the soil by gravity-drained water
II. dramatic fluctuations of the water table
III. an increased percentage of fines in the soil
IV. a decreased viscosity of the nonaqueous phase liquid

(A) I
(B) I and II
(C) III
(D) III and IV

Hint: Review nonaqueous phase liquid (NAPL) migration characteristics in soil.

PROBLEM 96

A well completely penetrates a confined aquifer under conditions shown in the following illustration. The hydraulic conductivity of the aquifer is 7.3 ft/day.

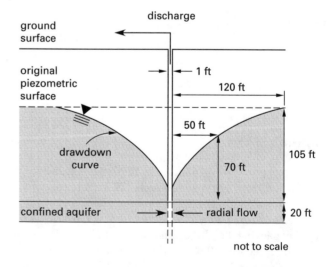

What is the flow from the well?

(A) 0.029 ft³/sec
(B) 0.16 ft³/sec
(C) 0.43 ft³/sec
(D) 1.9 ft³/sec

Hint: Discharge flow equations are available for confined and unconfined aquifers. Use the illustration to define terms.

PROBLEM 97

The extraction rate from an unconfined aquifer serving a large metropolitan area exceeds the recharge rate as demonstrated by the static water table's decline of 4.4 m over a 48 month monitoring period. The aquifer characteristics follow. Approximately how much water is lost from aquifer storage during the monitoring period?

aquifer horizontal surface area	512 km²
aquifer thickness	38 m
average porosity	0.43
hydraulic conductivity	0.38 cm/s
storativity	0.21

(A) 2.1×10^6 m³
(B) 4.7×10^8 m³
(C) 9.7×10^8 m³
(D) 3.6×10^9 m³

Hint: Review the definitions of the given aquifer characteristics.

PROBLEM 98

An agricultural drain field is depicted in the illustration. For a soil hydraulic conductivity of 0.018 cm/s and an average infiltration rate of 0.0014 cm/s, what is the required drain spacing?

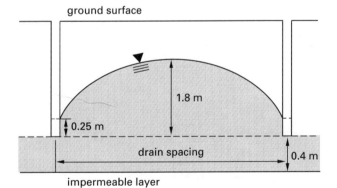

(A) 1.2 m
(B) 15 m
(C) 140 m
(D) 230 m

Hint: This is a classic drain spacing problem. Look for a drain spacing equation and be careful that the terms match those given in the illustration.

PROBLEM 99

Two 8 in diameter auger holes are drilled 10 ft apart through a 15 ft thick shallow unconfined aquifer into an underlying clay layer. Water is pumped from one hole to the other through a flow meter at 26 gal/min until the water level in each hole stabilizes with a head difference between the holes of 9.7 in. What is most nearly the hydraulic conductivity of the soil comprising the aquifer?

(A) 30 ft/day
(B) 130 ft/day
(C) 360 ft/day
(D) 620 ft/day

Hint: Look for equations describing the two-auger-hole method for determining hydraulic conductivity.

PROBLEM 100

Which of the compounds listed in the table is/are most likely to occur as dense nonaqueous phase liquids (DNAPL) in a soil-groundwater system?

compound	density (g/cm³)	solubility in water (mg/L at T°C)
acetone	0.79	infinite
methyl ethyl ketone	0.81	353 at 10
methylene chloride	1.33	16 700 at 25
naphthalene	1.03	32 at 25
tetrachloroethene	1.62	150 at 25
vinyl chloride	0.91	1.1 at 25

(A) acetone
(B) methyl ethyl ketone, vinyl chloride
(C) methylene chloride
(D) naphthalene, tetrachloroethene

Hint: Review the definition of dense nonaqueous phase liquids.

Breadth Solutions

WASTEWATER TREATMENT

SOLUTION 1

$$\text{VSS} = \frac{\text{MSS} - \text{MSI}}{\text{VF}}$$

$$= \left(\frac{25.645 \text{ g} - 25.501 \text{ g}}{200 \text{ mL}}\right)\left(\frac{10^6 \text{ mL·mg}}{\text{L·g}}\right)$$

$$= 720 \text{ mg/L}$$

The answer is (C).

Why Other Options Are Wrong

(A) This incorrect solution calculates the volatile total dissolved solids (volatile TDS).

$$\left(\frac{276.227 \text{ g} - 276.201 \text{ g}}{100 \text{ mL}}\right)\left(\frac{10^6 \text{ mL·mg}}{\text{L·g}}\right) = 260 \text{ mg/L}$$

(B) This incorrect solution calculates the nonvolatile (fixed) solids.

$$\left(\frac{25.501 \text{ g} - 25.439 \text{ g}}{200 \text{ mL}}\right)\left(\frac{10^6 \text{ mL·mg}}{\text{L·g}}\right) = 310 \text{ mg/L}$$

(D) This incorrect solution calculates the total suspended solids (TSS).

$$\left(\frac{25.645 \text{ g} - 25.439 \text{ g}}{200 \text{ mL}}\right)\left(\frac{10^6 \text{ mL·mg}}{\text{L·g}}\right)$$

$$= 1030 \text{ mg/L} \quad (1000 \text{ mg/L})$$

SOLUTION 2

E	fractional efficiency	–
Q	flow rate	m³/d
t	theoretical hydraulic detention time	d
t_a	actual hydraulic detention time	d
V	volume	m³

$$t = \frac{V}{Q} = \frac{(2.5 \text{ m})(15 \text{ m})(3.0 \text{ m})\left(24 \dfrac{\text{h}}{\text{d}}\right)}{900 \dfrac{\text{m}^3}{\text{d}}}$$

$$= 3.0 \text{ h}$$

$$t_a = tE$$

$$= (3.0 \text{ h})\left(\frac{83\%}{100\%}\right)$$

$$= 2.5 \text{ h}$$

The answer is (B).

Why Other Options Are Wrong

(A) This incorrect solution divides the flow rate by the volume to get the theoretical detention time. The theoretical detention time is then divided by the percent efficiency instead of being multiplied by the fractional efficiency. Other definitions and equations are unchanged from the correct solution.

$$t = \frac{Q}{V} = \frac{\left(900 \dfrac{\text{m}^3}{\text{d}}\right)\left(24 \dfrac{\text{h}}{\text{d}}\right)}{(2.5 \text{ m})(15 \text{ m})(3.0 \text{ m})}$$

$$= 192 \text{ h/d}^2$$

$$t_a = \frac{192 \dfrac{\text{h}}{\text{d}^2}}{83\%}$$

$$= 2.3 \text{ h}$$

Units do not make sense.

(C) This incorrect solution calculates the theoretical detention time. The hydraulic efficiency is ignored. Other definitions and equations are unchanged from the correct solution.

$$t = \frac{(2.5 \text{ m})(15 \text{ m})(3.0 \text{ m})\left(24 \dfrac{\text{h}}{\text{d}}\right)}{900 \dfrac{\text{m}^3}{\text{d}}}$$

$$= 3.0 \text{ h}$$

(D) In this incorrect solution, the theoretical detention time is divided by the hydraulic efficiency instead of being multiplied by it. Other definitions and equations are unchanged from the correct solution.

$$t = \frac{(2.5 \text{ m})(15 \text{ m})(3.0 \text{ m})\left(24 \dfrac{\text{h}}{\text{d}}\right)}{900 \dfrac{\text{m}^3}{\text{d}}}$$

$$= 3.0 \text{ h}$$

$$t_a = (3.0 \text{ h})\left(\frac{100\%}{83\%}\right)$$
$$= 3.6 \text{ h}$$

SOLUTION 3

The minimum standards for secondary wastewater treatment included in the CWA place effluent limits on 5 day biochemical oxygen demand (BOD_5), suspended solids, and hydrogen-ion concentration (pH). In some cases carbonaceous BOD_5 may be substituted for BOD_5. Although disinfection of effluents is required, the specific concerns of disinfection byproducts and dissolved solids are not included in the minimum standards for secondary treatment of wastewater, both being primarily associated with drinking water.

The answer is (A).

Why Other Options Are Wrong

(B) This response is incorrect because, although suspended solids are included, secondary wastewater treatment standards do not include disinfection byproducts.

(C) This response is incorrect because secondary wastewater treatment standards for solids are limited to suspended solids and do not include dissolved solids.

(D) This response is incorrect because neither disinfection byproducts nor dissolved solids are included among the minimum standards for secondary treatment of wastewater.

SOLUTION 4

time period (hr)	period average flow (10^6 gal/day)	period flow volume (10^6 gal)	cumulative volume (10^6 gal)
0000–0400	1.39	0.232	0.232
0400–0800	3.21	0.535	0.767
0800–1200	4.05	0.675	1.442
1200–1600	2.63	0.438	1.880
1600–2000	3.91	0.652	2.532
2000–2400	1.98	0.330	2.862
		2.862	

The time period represents a 4 hr interval.

Q period average flow rate 10^6 gal/day
t time period   hr
V flow volume 10^6 gal
V_c cumulative volume 10^6 gal

$$V = Qt = \frac{Q(4 \text{ hr})}{24 \frac{\text{hr}}{\text{day}}}$$

$$V_{c,i} = V_i + V_{c,i-1}$$

The daily average flow is

$$Q_a = \frac{2.862 \times 10^6 \text{ gal}}{6 \text{ periods}} = \frac{0.477 \times 10^6 \text{ gal}}{\text{period}}$$

The line plotted from the ordinate to the data point corresponding to 24 h represents the average flow. Lines representing the maximum deviation above and below the average flow are plotted parallel to the average flow line.

The volume represented by the deviation between the upper and lower parallel lines is the volume required for flow equalization as shown in the illustration.

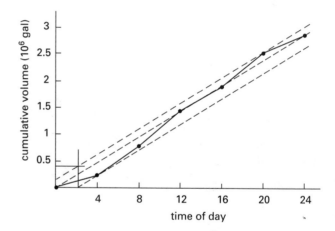

The tank volume, V_{tank}, is 5.0×10^5 gal.

The answer is (B).

Why Other Options Are Wrong

(A) This solution is incorrect because the required tank volume is calculated as the simple average of the total volume. Other definitions are unchanged from the correct solution.

V_T total flow volume   gal

$$V_T = 2.862 \times 10^6 \text{ gal}$$

$$V_{\text{tank}} = \frac{2.862 \times 10^6 \text{ gal}}{6}$$
$$= 4.8 \times 10^5 \text{ gal}$$

(C) This solution is incorrect because the required tank volume is determined using a plot of the cumulative average flow rate instead of the cumulative flow volume.

To get volume from the illustration, the cumulative flow units are incorrectly taken as gallons. Other definitions are unchanged from the correct solution.

time period (hr)	average flow (10^6 gal/day)	cumulative flow (10^6 gal/day)
0000–0400	1.39	1.39
0400–0800	3.21	4.60
0800–1200	4.05	8.65
1200–1600	2.63	11.3
1600–2000	3.91	12.2
2000–2400	1.98	17.2

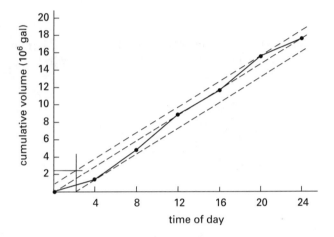

$$V_{\text{tank}} = 2.1 \times 10^6 \text{ gal}$$

(D) This solution is incorrect because the required tank volume is determined from the simple average of the sum of the average flows. Notice that this is the total flow volume. Other definitions are unchanged from the correct solution.

$$1.39 \times 10^6 \, \frac{\text{gal}}{\text{day}} + 3.21 \times 10^6 \, \frac{\text{gal}}{\text{day}} + 4.05 \times 10^6 \, \frac{\text{gal}}{\text{day}}$$
$$+ 2.63 \times 10^6 \, \frac{\text{gal}}{\text{day}} + 3.91 \times 10^6 \, \frac{\text{gal}}{\text{day}}$$
$$+ 1.98 \times 10^6 \, \frac{\text{gal}}{\text{day}}$$
$$= 17.2 \times 10^6 \text{ gal/day}$$

$$V_{\text{tank}} = \left(17.2 \times 10^6 \, \frac{\text{gal}}{\text{day}}\right) \left(\frac{1 \text{ day}}{24 \text{ hr}}\right) \left(\frac{4 \text{ hr}}{\text{period}}\right)$$
$$= 2.9 \times 10^6 \text{ gal}$$

SOLUTION 5

To find the recirculated solids flow rate, perform a mass balance around the clarifier based on biomass. A mass balance based on substrate will produce a trivial solution.

Q	flow rate	10^6 gal/day
Q_r	recirculated solids flow rate	10^6 gal/day
Q_w	wasted solids flow rate	10^6 gal/day
X	reactor mixed liquor suspended solids concentration	mg/L
X_e	clarifier effluent solids concentration	mg/L
X_u	recirculated solids concentration	mg/L

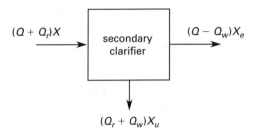

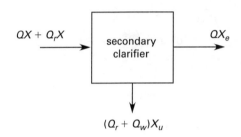

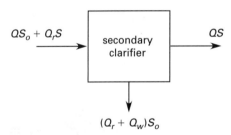

The mass of biomass equals the mass of biomass recirculated plus the mass of biomass wasted.

$$(Q + Q_r)X = (Q_r + Q_w)X_u + (Q - Q_w)X_e$$

Assume the clarifier effluent solids concentration is negligible compared to the reactor mixed liquor and recirculated solids concentrations.

$$X_e \ll X, X_u$$
$$(Q - Q_w)X_e \approx 0$$
$$(Q + Q_r)X = (Q_r + Q_w)X_u$$

Multiply through and reorder the equation in terms of the recirculated solids flow rate, Q_r.

$$QX + Q_rX = Q_rX_u + Q_wX_u$$
$$Q_r(X - X_u) = Q_wX_u - QX$$
$$Q_r = \frac{Q_wX_u - QX}{X - X_u}$$
$$= \frac{QX - Q_wX_u}{X_u - X}$$

All parameters are known except the wasted solids and recirculated solids flow rates. The wasted solids flow rate can be calculated from the definition of mean cell residence time.

V reactor volume 10^6 gal
θ_c mean cell residence time   day

$$\theta_c = \frac{VX}{Q_wX_u}$$
$$Q_w = \frac{VX}{\theta_cX_u}$$
$$= \frac{(5 \times 10^6 \text{ gal})\left(3500 \ \frac{\text{mg}}{\text{L}}\right)}{(10 \text{ day})\left(12\,000 \ \frac{\text{mg}}{\text{L}}\right)}$$
$$= 1.46 \times 10^5 \text{ gal/day}$$

$$Q_r = \frac{\left(2.5 \times 10^7 \ \frac{\text{gal}}{\text{day}}\right)\left(3500 \ \frac{\text{mg}}{\text{L}}\right) - \left(1.46 \times 10^5 \ \frac{\text{gal}}{\text{day}}\right)\left(12\,000 \ \frac{\text{mg}}{\text{L}}\right)}{12\,000 \ \frac{\text{mg}}{\text{L}} - 3500 \ \frac{\text{mg}}{\text{L}}}$$
$$= 1.0 \times 10^7 \text{ gal/day}$$

The answer is (A).

Why Other Options Are Wrong

(B) This incorrect solution is based on an improperly defined mass balance. The mass balance is performed around the clarifier using biomass, but the labeling of biomass inputs and outputs is incorrect. Other assumptions and definitions are unchanged from the correct solution.

$$QX + Q_rX_u = (Q_r + Q_w)X_u + QX_e$$
$$X_e \ll X, X_u$$
$$QX_e \approx 0$$
$$QX + Q_rX_u = (Q_r + Q_w)X_u$$
$$= Q_rX_u + Q_wX_u$$
$$QX = Q_wX_u$$
$$Q_w = \frac{QX}{X_u}$$
$$= \frac{\left(2.5 \times 10^7 \ \frac{\text{gal}}{\text{day}}\right)\left(3500 \ \frac{\text{mg}}{\text{L}}\right)}{12\,000 \ \frac{\text{mg}}{\text{L}}}$$
$$= 7.3 \times 10^6 \text{ gal/day}$$

Assume that the sum of the recirculated solids and wasted solids flow rates equal the influent flow rate.

$$Q_r = Q - Q_w$$
$$= 2.5 \times 10^7 \ \frac{\text{gal}}{\text{day}} - 7.3 \times 10^6 \ \frac{\text{gal}}{\text{day}}$$
$$= 1.77 \times 10^7 \text{ gal/day} \quad (1.8 \times 10^7 \text{ gal/day})$$

(C) This incorrect solution is based on an improperly defined mass balance. The mass balance is performed around the clarifier, but uses substrate instead of biomass and the substrate labeling is wrong. Other definitions and assumptions are unchanged from the correct solution.

S reactor substrate concentration   mg/L
S_o influent substrate concentration   mg/L

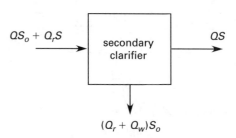

$$QS_o + Q_rS = (Q_r + Q_w)S_o + QS$$
$$= Q_rS_o + Q_wS_o + QS$$
$$Q(S_o - S) - Q_wS_o = Q_r(S_o - S)$$
$$Q_r = \frac{Q(S_o - S) - Q_wS_o}{S_o - S}$$

$$Q_w = \frac{VX}{\theta_c X_u} = \frac{(5.0 \times 10^6 \text{ gal})\left(3500 \frac{\text{mg}}{\text{L}}\right)}{(10 \text{ day})\left(12\,000 \frac{\text{mg}}{\text{L}}\right)}$$

$$= 1.46 \times 10^5 \text{ gal/day}$$

$$Q_r = \frac{\left(2.5 \times 10^7 \frac{\text{gal}}{\text{day}}\right)\left(224 \frac{\text{mg}}{\text{L}} - 20 \frac{\text{mg}}{\text{L}}\right)}{224 \frac{\text{mg}}{\text{L}} - 20 \frac{\text{mg}}{\text{L}}}$$

$$ = \frac{-\left(1.46 \times 10^5 \frac{\text{gal}}{\text{day}}\right)\left(224 \frac{\text{mg}}{\text{L}}\right)}{224 \frac{\text{mg}}{\text{L}} - 20 \frac{\text{mg}}{\text{L}}}$$

$$= 2.5 \times 10^7 \text{ gal/day}$$

(D) This incorrect solution applies the "plug-flow" equation for activated sludge solved for recirculation ratio. Other assumptions and definitions are unchanged from the correct solution.

Assume typical values for the following.

k	growth rate constant	5 day^{-1}
K_s	half velocity constant	60 mg/L
SF	safety factor	10
θ_c^d	design mean cell residence time	day
θ_c^m	minimum mean cell residence time	day

$$\theta_c^m = \frac{\theta_c^d}{SF} = \frac{10 \text{ day}}{10}$$

$$= 1 \text{ day}$$

k_d	endogenous decay rate constant	day^{-1}
R	recirculation ratio	–
S_i	reactor substrate concentration after mixing with recirculated flow	mg/L
Y	yield coefficient	g/g

$$\frac{1}{\theta_c^m} = \frac{Yk(S_o - S)}{(S_o - S) + (1 + R)K_s \ln\left(\frac{S_i}{S}\right)} - k_d$$

$$S_i = \frac{S_o + RS}{1 + R}$$

$$\frac{1}{1 \text{ day}} = \frac{\left(0.5 \frac{\text{g}}{\text{g}}\right)(5 \text{ day}^{-1})\left(224 \frac{\text{mg}}{\text{L}} - 20 \frac{\text{mg}}{\text{L}}\right)}{\left(224 \frac{\text{mg}}{\text{L}} - 20 \frac{\text{mg}}{\text{L}}\right) + (1+R)\left(60 \frac{\text{mg}}{\text{L}}\right)}$$

$$\times \ln\left(\frac{224 \frac{\text{mg}}{\text{L}} + R\left(20 \frac{\text{mg}}{\text{L}}\right)}{\left(20 \frac{\text{mg}}{\text{L}}\right)(1 + R)}\right)$$

$$- 0.05 \text{ day}^{-1}$$

Cancel units and simplify.

$$1.05 \text{ day}^{-1} = \frac{510 \frac{\text{mg}}{\text{L} \cdot \text{day}}}{204 \frac{\text{mg}}{\text{L}} + (1+R)\left(60 \frac{\text{mg}}{\text{L}}\right)}$$

$$\times \ln\left(\frac{11.2 + R}{1 + R}\right)$$

$$4.7 = (1+R)\ln\left(\frac{11.2 + R}{1 + R}\right)$$

All units cancel.

Solve for the recirculation ratio by trial and error.

$$R = 2.4$$

$$Q_r = RQ = (2.4)\left(2.5 \times 10^7 \frac{\text{gal}}{\text{day}}\right)$$

$$= 6.0 \times 10^7 \text{ gal/day}$$

SOLUTION 6

f	solids fraction	–
$\dot{m}$	solids dry mass flow rate	lbm/day
$\dot{V}$	volumetric flow rate	gal/day
ρ	wet sludge density	lbm/ft^3

$$\dot{m} = f_1 \rho \dot{V}_1 = f_2 \rho \dot{V}_2$$

$$\dot{V}_2 = \frac{f_1 \dot{V}_1}{f_2}$$

At 1.2% solids,

$$\dot{V}_1 = 50{,}000 \text{ gal/day}$$

At 24% solids,

$$\dot{V}_2 = \frac{(0.012)\left(50{,}000 \frac{\text{gal}}{\text{day}}\right)}{0.24}$$

$$= 2500 \text{ gal/day}$$

$\dot{V}_r$ volume flow rate reduction realized   gal/day

$$\dot{V}_r = \dot{V}_1 - \dot{V}_2$$

$$= 50{,}000 \frac{\text{gal}}{\text{day}} - 2500 \frac{\text{gal}}{\text{day}}$$

$$= 47{,}500 \text{ gal/day} \quad (48{,}000 \text{ gal/day})$$

The answer is (D).

Why Other Options Are Wrong

(A) This incorrect choice is the calculated sludge volume flow rate at 24% solids. The difference between the volume flow rate at 24% solids and 1.2% solids is required for the correct answer. Other equations and definitions are the same as used in the correct solution.

Assume the solids density to be approximately equal to that of water at 8.342 lbm/gal.

At 1.2% solids,

$$\dot{V} = 50{,}000 \text{ gal/day}$$

$$\dot{m} = (0.012)\left(50{,}000 \ \frac{\text{gal}}{\text{day}}\right)\left(8.342 \ \frac{\text{lbm}}{\text{gal}}\right)$$

$$= 5005 \text{ lbm/day}$$

At 24% solids,

$$\dot{m} = 5005 \text{ lbm/day}$$

$$\dot{V}_r = \frac{\dot{m}}{f_2 \rho} = \frac{5005 \ \frac{\text{lbm}}{\text{day}}}{(0.24)\left(8.342 \ \frac{\text{lbm}}{\text{gal}}\right)}$$

$$= 2500 \text{ gal/day}$$

(B) This incorrect choice uses the difference between 24% and 1.2% solids and solids mass, and fails to use the mass/density/volume relationship consistently. Other equations and definitions are the same as used in the correct solution.

Assume the solids density to be approximately equal to that of water at 8.342 lbm/gal.

At 1.2% solids,

$$\dot{V} = 50{,}000 \text{ gal/day}$$

$$\dot{m} = (0.012)\left(50{,}000 \ \frac{\text{gal}}{\text{day}}\right)\left(8.342 \ \frac{\text{lbm}}{\text{gal}}\right)$$

$$= 5005 \text{ lbm/day}$$

$$\dot{V}_r = \frac{\dot{m}}{(f_2 - f_1)\rho} = \frac{5005 \ \frac{\text{lbm}}{\text{day}}}{(0.24 - 0.012)\left(8.342 \ \frac{\text{lbm}}{\text{gal}}\right)}$$

$$= 2630 \text{ gal/day} \quad (2600 \text{ gal/day})$$

(C) This incorrect choice does not conserve mass. Other equations and definitions are the same as used in the correct solution.

At 1.2% solids,

$$\dot{V}_1 = (f_t - f_1)\dot{V}_1 = 50{,}000 \text{ gal/day}$$

At 24% solids,

$$\dot{V}_2 = (0.24 - 0.012)\left(50{,}000 \ \frac{\text{gal}}{\text{day}}\right)$$

$$= 11{,}400 \text{ gal/day}$$

$$\dot{V}_r = 50{,}000 \ \frac{\text{gal}}{\text{day}} - 11{,}400 \ \frac{\text{gal}}{\text{day}}$$

$$= 38{,}600 \text{ gal/day} \quad (39{,}000 \text{ gal/day})$$

AQUATIC BIOLOGY AND MICROBIOLOGY

SOLUTION 7

Bacteria and viruses are small organisms that, unless flocculated or sorbed to larger particles, will pass through a filter bed. These are commonly targeted in application of chlorine for disinfection and are routinely analyzed using indicator organisms and other methods. Protozoa include *Giardia lamblia* and *cryptosporidium* and helminths include pinworms, roundworms, and tapeworms. Protozoan oocysts protect the organisms during chlorination. Although less resistant to the effects of chlorination than protozoa, helminths and helminth ova along with protozoa are primarily removed by filtration. Where filtration is not practiced as part of wastewater treatment, large populations of helminths and protozoa may be discharged to receiving waters.

The answer is (D).

Why Other Options Are Wrong

(A) This option is incorrect because both protozoa and helminths are infrequently looked for in routine analysis, are generally less susceptible to chlorination, and are more likely to be targeted for removal by filtration.

(B) This option is incorrect because both viruses and bacteria are readily susceptible to chlorination, are the object of routine analyses, and, unless sorbed to other particles or flocculated, are not targeted by filtration.

(C) This option is incorrect because viruses are not generally resistant to chlorination or effectively removed by filtration, unless sorbed to larger particles or flocculated. Helminths are not included in routine analyses and are removed by filtration. Also, helminths may be less affected by chlorination than viruses are.

SOLUTION 8

Because they are routinely cultured in the laboratory, it is desirable that indicator organisms not be pathogenic. Although some coliform organisms are infectious, they are generally regarded as nonpathogenic.

The answer is (D).

Why Other Options Are Wrong

(A) This choice is incorrect because indicator organisms apply to all types of water. This is necessary for reliable results.

(B) This choice is incorrect because indicator organisms are only present in large numbers when pathogens are present and are generally absent, or present in small numbers, otherwise. This is necessary to prevent false positive and false negative results.

(C) This choice is incorrect because indicator organisms are easily detected using routine analytical procedures. This is necessary to allow rapid analysis and minimum cost.

SOLUTION 9

Typically bottles outside the range of 2.0 mg/L to 7.0 mg/L final dissolved oxygen (DO) are excluded from calculations for biochemical oxygen demand (BOD). Disregard bottles 1 and 4 since the final DO for these bottles is outside of the typically accepted range of 2.0 mg/L to 7.0 mg/L.

$DO_{t=0}$	dissolved oxygen concentration at $t = 0$ d	mg/L
$DO_{t=5}$	dissolved oxygen concentration at $t = 5$ d	mg/L
t	time	d
V	volume	mL

For each bottle,

$$\frac{DO_{t=0} - DO_{t=5}}{\frac{\text{sample } V}{\text{total } V}}$$

Assume a typical BOD bottle total volume is 300 mL.

For bottle 2,

$$\frac{9.1 \frac{\text{mg}}{\text{L}} - 2.9 \frac{\text{mg}}{\text{L}}}{\frac{10 \text{ mL}}{300 \text{ mL}}} = 186 \text{ mg/L}$$

For bottle 3,

$$\frac{9.1 \frac{\text{mg}}{\text{L}} - 6.1 \frac{\text{mg}}{\text{L}}}{\frac{5 \text{ mL}}{300 \text{ mL}}} = 180 \text{ mg/L}$$

The BOD_5 at 20°C is

$$\frac{186 \frac{\text{mg}}{\text{L}} + 180 \frac{\text{mg}}{\text{L}}}{2} = 183 \text{ mg/L}$$

k	rate constant	d^{-1}
BOD_u	ultimate BOD	mg/L
BOD_t	BOD at time t	mg/L

$$BOD_u = \frac{BOD_t}{1 - 10^{-kt}}$$

$$= \frac{183 \frac{\text{mg}}{\text{L}}}{1 - 10^{(-0.17/\text{d})(5 \text{ d})}}$$

$$= 213 \text{ mg/L}$$

k_T	rate constant at temperature of interest	d^{-1}
k_{20}	rate constant at 20°C	d^{-1}
T	temperature	°C
θ	temperature correction coefficient	–

Assume a typical value of 1.047 for the temperature correction coefficient.

$$k_T = k_{20}\theta^{T-20}$$

$$k_{15} = \left(\frac{0.17}{\text{d}}\right)(1.047^{15-20}) = 0.135 \text{ d}^{-1}$$

The BOD_5 at 15°C is

$$\left(213 \frac{\text{mg}}{\text{L}}\right)\left(1 - 10^{(-0.135/\text{d})(5 \text{ d})}\right)$$
$$= 168 \text{ mg/L} \quad (170 \text{ mg/L})$$

The answer is (C).

Why Other Options Are Wrong

(A) This incorrect solution improperly applies the ultimate BOD equation. Other assumptions, definitions, and equations are unchanged from the correct solution.

For bottle 2,

$$\frac{9.1 \frac{\text{mg}}{\text{L}} - 2.9 \frac{\text{mg}}{\text{L}}}{\frac{10 \text{ mL}}{300 \text{ mL}}} = 186 \text{ mg/L}$$

For bottle 3,

$$\frac{9.1 \frac{\text{mg}}{\text{L}} - 6.1 \frac{\text{mg}}{\text{L}}}{\frac{5 \text{ mL}}{300 \text{ mL}}} = 180 \text{ mg/L}$$

The BOD_5 at 20°C is

$$\frac{186 \frac{\text{mg}}{\text{L}} + 180 \frac{\text{mg}}{\text{L}}}{2} = 183 \text{ mg/L}$$

$$k_{15} = \left(\frac{0.17}{\text{d}}\right)(1.047^{15-20}) = 0.135 \text{ d}^{-1}$$

$$BOD_5 \text{ at } 15°C = BOD_5 \text{ at } 20°C \left(1 - 10^{(-k_{15})(t)}\right)$$
$$= \left(183 \frac{\text{mg}}{\text{L}}\right)\left(1 - 10^{(-0.135/\text{d})(5 \text{ d})}\right)$$
$$= 144 \frac{\text{mg}}{\text{L}} \quad \left(140 \frac{\text{mg}}{\text{L}}\right)$$

(B) This incorrect solution did not disregard bottles 1 and 4. Other assumptions, definitions, and equations were unchanged from the correct solution.

For bottle 1,

$$\frac{9.0 \ \frac{mg}{L} - 1.0 \ \frac{mg}{L}}{\frac{20 \ mL}{300 \ mL}} = 120 \ mg/L$$

For bottle 2,

$$\frac{9.1 \ \frac{mg}{L} - 2.9 \ \frac{mg}{L}}{\frac{10 \ mL}{300 \ mL}} = 186 \ mg/L$$

For bottle 3,

$$\frac{9.1 \ \frac{mg}{L} - 6.1 \ \frac{mg}{L}}{\frac{5 \ mL}{300 \ mL}} = 180 \ mg/L$$

For bottle 4,

$$\frac{9.2 \ \frac{mg}{L} - 8.1 \ \frac{mg}{L}}{\frac{2 \ mL}{300 \ mL}} = 165 \ mg/L$$

The BOD_5 at 20°C is

$$\frac{120 \ \frac{mg}{L} + 186 \ \frac{mg}{L} + 180 \ \frac{mg}{L} + 165 \ \frac{mg}{L}}{4} = 163 \ mg/L$$

$$BOD_u = \frac{163 \ \frac{mg}{L}}{1 - 10^{(-0.17/d)(5 \ d)}}$$
$$= 190 \ mg/L$$

$$k_{15} = \left(\frac{0.17}{d}\right)(1.047^{15-20})$$
$$= 0.135 \ d^{-1}$$

The BOD_5 at 15°C is

$$(190 \ mg/L)\left(1 - 10^{(-0.135/d)(5 \ d)}\right) = 150 \ mg/L$$

(D) This incorrect solution does not correct for temperature. Other assumptions, definitions, and equations are unchanged from the correct solution.

For bottle 2,

$$\frac{9.1 \ \frac{mg}{L} - 2.9 \ \frac{mg}{L}}{\frac{10 \ mL}{300 \ mL}} = 186 \ mg/L$$

For bottle 3,

$$\frac{9.1 \ \frac{mg}{L} - 6.1 \ \frac{mg}{L}}{\frac{5 \ mL}{300 \ mL}} = 180 \ mg/L$$

$$BOD_5 = \frac{186 \ \frac{mg}{L} + 180 \ \frac{mg}{L}}{2}$$
$$= 183 \ mg/L \quad (180 \ mg/L)$$

SOLUTION 10

At the discharge point, assume influences on dissolved oxygen (DO) from reaeration and microbial activity are negligible.

A mass balance diagram shows the inflow and outflow rates and concentrations.

d	downstream	–
DO	dissolved oxygen concentration	mg/L
$\dot{m}$	mass flow rate of dissolved oxygen	mg/s
Q	flow ft³/sec or gal/day	
u	upstream	–
W	wastewater treatment plant	–

wastewater treatment plant
$Q_W = 15 \times 10^6$ gal/day

upstream
$Q_u = 2000$ ft³/sec
$DO_u = 8.10$ mg/L

downstream
Q_d
DO_d

$$Q_d = Q_u + Q_W$$
$$= \left(2000 \ \frac{ft^3}{sec}\right)\left(28.25 \ \frac{L}{ft^3}\right) + \left(15 \times 10^6 \ \frac{gal}{day}\right)$$
$$\times \left(3.785 \ \frac{L}{gal}\right)\left(\frac{1 \ day}{86,400 \ sec}\right)$$
$$= 56\,500 \ \frac{L}{s} + 657 \ \frac{L}{s} = 57\,157 \ L/s$$

$$\dot{m}_d = \dot{m}_u + \dot{m}_W$$
$$= \left(56\,500 \ \frac{L}{s}\right)\left(8.10 \ \frac{mg}{L}\right) + \left(657 \ \frac{L}{s}\right)$$
$$\times \left(1.20 \ \frac{mg}{L}\right)$$
$$= 458\,438 \ mg/s$$

$$DO_d = \frac{\dot{m}_d}{Q_d} = \frac{458\,438 \ \frac{mg}{s}}{57\,157 \ \frac{L}{s}}$$
$$= 8.02 \ mg/L$$

The answer is (B).

Why Other Options Are Wrong

(A) This incorrect solution uses the average of the upstream and WWTP dissolved oxygen concentrations to compute the downstream concentration. The figure and other definitions are unchanged from the correct solution.

$$DO_d = \frac{DO_u + DO_W}{2}$$
$$= \frac{8.10\ \frac{mg}{L} + 1.20\ \frac{mg}{L}}{2}$$
$$= 4.65\ mg/L$$

(C) This incorrect solution does not convert flow rates to common units. The figure and other definitions are unchanged from the correct solution.

$$Q_d = Q_u + Q_W$$
$$= 2000\ \frac{ft^3}{sec} + 15 \times 10^6\ \frac{gal}{day}$$
$$= 2015$$

Units cannot be added and must be ignored.

$$\dot{m}_d = \dot{m}_u + \dot{m}_W$$
$$= \left(2000\ \frac{ft^3}{sec}\right)\left(8.10\ \frac{mg}{L}\right)$$
$$+ \left(15 \times 10^6\ \frac{gal}{day}\right)\left(1.20\ \frac{mg}{L}\right)$$
$$= 16{,}218$$

Units do not make sense and must be ignored.

$$DO_d = \frac{\dot{m}_d}{Q_d} = \frac{16{,}218}{2015}$$
$$= 8.05\ mg/L$$

(D) This incorrect solution improperly applies the mass balance. The figure and other definitions are unchanged from the correct solution.

$$Q_d = Q_u - Q_W$$
$$= \left(2000\ \frac{ft^3}{sec}\right)\left(28.25\ \frac{L}{ft^3}\right)$$
$$- \left(15 \times 10^6\ \frac{gal}{day}\right)\left(3.785\ \frac{L}{gal}\right)$$
$$\times \left(\frac{1\ day}{86{,}400\ sec}\right)$$
$$= 56\,500\ \frac{L}{s} - 657\ \frac{L}{s}$$
$$= 55\,843\ L/s$$

$$\dot{m}_d = \dot{m}_u - \dot{m}_W$$
$$= \left(56\,500\ \frac{L}{s}\right)\left(8.10\ \frac{mg}{L}\right)$$
$$- \left(657\ \frac{L}{s}\right)\left(1.20\ \frac{mg}{L}\right)$$
$$= 456\,862\ mg/s$$
$$DO_d = \frac{\dot{m}_d}{Q_d} = \frac{456\,862\ mg/s}{55\,843\ L/s}$$
$$= 8.18\ mg/L$$

SOLUTION 11

The EPA-recommended exposure factors are

BW	body weight	70 kg
DI	daily intake	2 L/d
ED	exposed duration	30 yr
LT	lifetime	70 yr
A	absorbed	(assume)
C	concentration	mg/L
CDI	chronic daily intake	mg/kg·d
PF	potency factor	(mg/kg·d)$^{-1}$
T	time exposed	(assume)

$$CDI = \frac{C(DI)(A)(ED)(T)}{(BW)(LT)}$$

For trichloroethylene (TCE),

$$C = \left(100\ \frac{\mu g}{L}\right)\left(1 \times 10^{-3}\ \frac{mg}{\mu g}\right)$$
$$= 0.1\ \frac{mg}{L}$$

$$CDI = \frac{\left(0.1\ \frac{mg}{L}\right)\left(2\ \frac{L}{d}\right)(1)(30\ yr)(1)}{(70\ kg)(70\ yr)}$$
$$= 0.001\,22\ mg/kg\cdot d$$

The lifetime risk is

$$(CDI)(PF_{TCE}) = \left(0.001\,22\ \frac{mg}{kg\cdot d}\right)\left(0.011\ \frac{1}{\frac{mg}{kg\cdot d}}\right)$$
$$= 1.3 \times 10^{-5}$$

For 1,1-dichloroethylene (1,1-DCE),

$$C = 7.2\ ppb$$
$$= 7.2\ \frac{\mu g}{L}\left(1 \times 10^{-3}\ \frac{mg}{\mu g}\right)$$
$$= 0.0072\ mg/L$$

$$\text{CDI} = \frac{\left(0.0072\ \frac{\text{mg}}{\text{L}}\right)\left(2\ \frac{\text{L}}{\text{d}}\right)(1)(30\ \text{yr})(1)}{(70\ \text{kg})(70\ \text{yr})}$$
$$= 0.000\,088\ \text{mg/kg·d}$$

The lifetime risk is

$$(\text{CDI})(\text{PF}_{1,1\text{-DCE}}) = \left(0.000\,088\ \frac{\text{mg}}{\text{kg·d}}\right)\left(0.58\ \frac{1}{\frac{\text{mg}}{\text{kg·d}}}\right)$$
$$= 5.1 \times 10^{-5}$$

The total lifetime risk is

$$1.3 \times 10^{-5} + 5.1 \times 10^{-5} = 6.4 \times 10^{-5} \quad (64\ \text{in one million})$$

The answer is (B).

Why Other Options Are Wrong

(A) This incorrect solution averages, instead of sums, individual risks. Other assumptions, definitions, and equations are the same as the correct solution.

For TCE,

$$C = \left(100\ \frac{\mu\text{g}}{\text{L}}\right)\left(1 \times 10^{-3}\ \frac{\text{mg}}{\mu\text{g}}\right)$$
$$= 0.1\ \text{mg/L}$$

$$\text{CDI} = \frac{\left(0.1\ \frac{\text{mg}}{\text{L}}\right)\left(2\ \frac{\text{L}}{\text{d}}\right)(1)(30\ \text{yr})(1)}{(70\ \text{kg})(70\ \text{yr})}$$
$$= 0.001\,22\ \text{mg/kg·d}$$

The lifetime risk is

$$(\text{CDI})(\text{PF}_{\text{TCE}}) = \left(0.001\,22\ \frac{\text{mg}}{\text{kg·d}}\right)(0.011)\left(\frac{\text{mg}}{\text{kg·d}}\right)^{-1}$$
$$= 1.3 \times 10^{-5}$$

For 1,1-DCE,

$$C = 7.2\ \text{ppb}$$
$$= 7.2\ \frac{\mu\text{g}}{\text{L}}$$
$$= 0.0072\ \text{mg/L}$$

$$\text{CDI} = \frac{\left(0.0072\ \frac{\text{mg}}{\text{L}}\right)\left(2\ \frac{\text{L}}{\text{d}}\right)(1)(30\ \text{yr})(1)}{(70\ \text{kg})(70\ \text{yr})}$$
$$= 0.000\,088\ \text{mg/kg·d}$$

The lifetime risk is

$$(\text{CDI})(\text{PF}_{1,1\text{-DCE}}) = \left(0.000\,088\ \frac{\text{mg}}{\text{kg·d}}\right)(0.58)$$
$$\times \left(\frac{\text{mg}}{\text{kg·d}}\right)^{-1}$$
$$= 5.1 \times 10^{-5}$$

The total lifetime risk is

$$\frac{1.3 \times 10^{-5} + 5.1 \times 10^{-5}}{2}$$
$$= 3.2 \times 10^{-5} \quad (32\ \text{in one million})$$

(C) This incorrect solution uses 70 yr for exposed duration and lifetime, and averages individual risks. Other assumptions, definitions, and equations are the same as the correct solution.

The EPA-recommended exposure factors are

DI   2 L/d
ED   70 yr
BW   70 kg
LT   70 yr

For TCE,

$$C = \left(100\ \frac{\mu\text{g}}{\text{L}}\right)\left(1 \times 10^{-3}\ \frac{\text{mg}}{\mu\text{g}}\right)$$
$$= 0.1\ \text{mg/L}$$

$$\text{CDI} = \frac{\left(0.1\ \frac{\text{mg}}{\text{L}}\right)\left(2\ \frac{\text{L}}{\text{d}}\right)(1)(70\ \text{yr})(1)}{(70\ \text{kg})(70\ \text{yr})}$$
$$= 0.002\,86\ \text{mg/kg·d}$$

The lifetime risk is

$$(\text{CDI})(\text{PF}_{\text{TCE}}) = \left(0.002\,86\ \frac{\text{mg}}{\text{kg·d}}\right)(0.011)$$
$$\times \left(\frac{\text{mg}}{\text{kg·d}}\right)^{-1}$$
$$= 3.1 \times 10^{-5}$$

For 1,1-DCE,

$$C = \left(7.2\ \frac{\mu\text{g}}{\text{L}}\right)\left(1 \times 10^{-3}\ \frac{\text{mg}}{\mu\text{g}}\right)$$
$$= 0.0072\ \text{mg/L}$$

$$\text{CDI} = \frac{\left(0.0072\ \frac{\text{mg}}{\text{L}}\right)\left(2\ \frac{\text{L}}{\text{d}}\right)(1)(70\ \text{yr})(1)}{(70\ \text{kg})(70\ \text{yr})}$$
$$= 0.000\,206\ \text{mg/kg·d}$$

The lifetime risk is

$$(\text{CDI})(\text{PF}_{1,1\text{-DCE}}) = \left(0.000\,206\ \frac{\text{mg}}{\text{kg·d}}\right)(0.58)$$
$$\times \left(\frac{\text{mg}}{\text{kg·d}}\right)^{-1}$$
$$= 1.2 \times 10^{-4}$$

The total lifetime risk is

$$\frac{3.1 \times 10^{-5} + 1.2 \times 10^{-4}}{2}$$
$$= 7.6 \times 10^{-5} \quad (76\ \text{in one million})$$

(D) This incorrect solution uses the summed concentration and the summed potency factor to calculate risk. Other assumptions, definitions, and equations are the same as the correct solution.

$$C = 100 \,\frac{\mu g}{L} + 7.2 \,\frac{\mu g}{L} = 107.2 \,\frac{\mu g}{L}$$

$$= \left(107.2 \,\frac{\mu g}{L}\right)\left(1 \times 10^{-3} \,\frac{\mu g}{L}\right)$$

$$= 0.1072 \text{ mg/L}$$

$$\text{CDI} = \frac{\left(0.1072 \,\frac{mg}{L}\right)\left(2 \,\frac{L}{d}\right)(1)(30 \text{ yr})(1)}{(70 \text{ kg})(70 \text{ yr})}$$

$$= 0.0013 \text{ mg/kg·d}$$

$$\text{PF} = 0.011 \left(\frac{mg}{kg \cdot d}\right)^{-1} + 0.58 \left(\frac{mg}{kg \cdot d}\right)^{-1}$$

$$= 0.591 \left(\frac{mg}{kg \cdot d}\right)^{-1}$$

The total lifetime risk is

$$(\text{CDI})(\text{PF}) = \left(0.0013 \,\frac{mg}{kg \cdot d}\right)(0.591)$$

$$\times \left(\frac{mg}{kg \cdot d}\right)^{-1}$$

$$= 7.7 \times 10^{-4} \quad (770 \text{ in one million})$$

SOLID AND HAZARDOUS WASTE

SOLUTION 12

Assume a 100 kg sample of the waste for convenience in calculations. Although the number of residents and the generation rate can be used to calculate the total daily mass, it will not affect the waste density result.

component	component discarded mass (kg)	component discarded density (kg/m³)	component discarded volume (m³)
paper	31	85	0.365
garden	29	105	0.276
food	10	290	0.034
cardboard	9	50	0.180
wood	8	240	0.033
plastic	7	65	0.108
miscellaneous	6	480	0.013
	100		1.00

m	component discarded mass	kg
V	component discarded volume	m³
ρ	waste bulk wet density	kg/m³
ρ_d	component discarded density	kg/m³

$$V = \frac{m}{\rho_d}$$

$$\rho = \frac{\sum m}{\sum V} = \frac{100 \text{ kg}}{1.00 \text{ m}^3}$$

$$= 100 \text{ kg/m}^3$$

The answer is (B).

Why Other Options Are Wrong

(A) This incorrect solution calculates dry density instead of wet density. Other assumptions, definitions, and equations are unchanged from the correct solution.

m_o	component dry mass	kg
V_o	component dry volume	m³
W_d	component discarded moisture	kg
$W\%$	component discarded moisture	%

Table for Solution 12, Option (A)

component	component discarded mass (kg)	component discarded moisture (kg)	component dry mass (kg)	component discarded density (kg/m³)	component dry volume (m³)
paper	31	1.86	29.1	85	0.342
garden	29	17.4	11.6	105	0.110
food	10	7.0	3.00	290	0.010
cardboard	9	0.45	8.55	50	0.171
wood	8	1.6	6.40	240	0.027
plastic	7	0.14	6.86	65	0.106
miscellaneous	6	0.48	5.52	480	0.012
	100	28.9	71.0		0.78

$$W_d = \frac{mW\%}{100\%}$$
$$m_o = m - W_d$$
$$V_o = \frac{m_o}{\rho_d}$$
$$\rho = \frac{m_o}{V_o}$$
$$= \frac{71 \text{ kg}}{0.78 \text{ m}^3}$$
$$= 91 \text{ kg/m}^3$$

(C) This incorrect solution calculates density using the sum of the mass fractions of the discarded densities. Other assumptions, definitions, and equations are unchanged from the correct solution.

component	mass (%)	component discarded density (kg/m³)	component fractional density (kg/m³)
paper	31	85	26.4
garden	29	105	30.5
food	10	290	29.0
cardboard	9	50	4.5
wood	8	240	19.2
plastic	7	65	4.55
miscellaneous	6	480	28.8
			143

ρ_f component fractional density   kg/m³

$$\rho_f = \frac{\rho_d(\% \text{ mass})}{100\%}$$
$$\rho = \sum \rho_f$$
$$= 143 \text{ kg/m}^3 \quad (140 \text{ kg/m}^3)$$

(D) This incorrect solution calculates the density using the sum of the component discarded densities. Other definitions are unchanged from the correct solution.

component	component discarded density (kg/m³)
paper	85
garden	105
food	290
cardboard	50
wood	240
plastic	65
miscellaneous	480
	1315

$$\rho = 1315 \text{ kg/m}^3 \quad (1300 \text{ kg/m}^3)$$

SOLUTION 13

Assume that pick-up occurs once weekly.

The residences per truck volume are

$$(20 \text{ yd}^3)\left(\frac{1 \text{ residence}}{0.29 \text{ yd}^3}\right) = 69 \text{ residences}$$

The time per residence is

$$(14 \text{ sec} + 32 \text{ sec})\left(\frac{1 \text{ min}}{60 \text{ sec}}\right) = 0.77 \text{ min}$$

The time to fill one truck is

$$(69 \text{ residences})\left(0.77 \frac{\text{min}}{\text{residence}}\right) = 53 \text{ min}$$

Assume that the time available for collection is one 8 hr day.

The total available minutes in a single day is

$$\left(8 \frac{\text{hr}}{\text{day}}\right)\left(60 \frac{\text{min}}{\text{hr}}\right) = 480 \text{ min}$$

activity	task time (min)	remaining time (min)
yard to route	25	455
collection 1	53	402
route to transfer	45	357
unload	15	342
transfer to route	45	297
collection 2	53	244
route to transfer	45	199
unload	15	184
transfer to route	45	139
collection 3	53	86
route to transfer	45	41
unload	15	26
transfer to yard	25	1

In a single day, one truck can complete three collection cycles. One truck and crew can service

$$\left(3 \frac{\text{collection cycles}}{\text{day}}\right)\left(69 \frac{\text{residences}}{\text{collection cycle}}\right)$$
$$= 207 \text{ residences/day} \quad (210 \text{ residences/day})$$

The answer is (B).

Why Other Options Are Wrong

(A) This incorrect solution uses the truck compacted waste capacity instead of the equivalent noncompacted capacity. Other assumptions are unchanged from the correct solution.

The residences per truck volume are

$$(8 \text{ yd}^3) \left(\frac{1 \text{ residence}}{0.29 \text{ yd}^3} \right) = 28 \text{ residences}$$

The time per residence is

$$(14 \text{ sec} + 32 \text{ sec}) \left(\frac{1 \text{ min}}{60 \text{ sec}} \right) = 0.77 \text{ min}$$

The time required to fill one truck is

$$(28 \text{ residences})(0.77 \text{ min}) = 22 \text{ min}$$

activity	task time (min)	remaining time (min)
yard to route	25	455
collection 1	22	433
route to transfer	45	388
unload	15	373
transfer to route	45	328
collection 2	22	306
route to transfer	45	261
unload	15	246
transfer to route	45	201
collection 3	22	179
route to transfer	45	134
unload	15	119
transfer to route	45	74
collection 4	22	52
route to yard	25	27

Assume that the truck unloads at the beginning of the next day.

One truck and crew can complete four collection cycles in a single day and can service

$$\left(\frac{4 \text{ collection cycles}}{\text{day}} \right) \left(\frac{28 \text{ residences}}{\text{collection cycle}} \right)$$
$$= 112 \text{ residences/day} \quad (110 \text{ residences/day})$$

(C) This incorrect solution does not include all travel and unloading times. Other assumptions are unchanged from the correct solution.

The residences per truck volume are

$$(20 \text{ yd}^3) \left(\frac{1 \text{ residence}}{0.29 \text{ yd}^3} \right) = 69 \text{ residences}$$

The time at each residence is

$$(14 \text{ sec} + 32 \text{ sec}) \left(\frac{1 \text{ min}}{60 \text{ sec}} \right) = 0.77 \text{ min}$$

The time required to fill one truck is

$$(69 \text{ residences})(0.77 \text{ min}) = 53 \text{ min}$$

The time not used for collection is

$$25 \text{ min} + 45 \text{ min} + 15 \text{ min} + 25 \text{ min} = 110 \text{ min}$$

One truck and crew can service

$$\left(480 \frac{\text{min}}{\text{day}} - 110 \frac{\text{min}}{\text{day}} \right) \left(\frac{69 \text{ residences}}{53 \text{ min}} \right)$$
$$= 482 \text{ residences/day} \quad (480 \text{ residences/day})$$

(D) This incorrect solution does not include travel and unloading times. Other assumptions are unchanged from the correct solution.

The residences per truck volume are

$$(20 \text{ yd}^3) \left(\frac{1 \text{ residence}}{0.29 \text{ yd}^3} \right) = 69 \text{ residences}$$

The time at each residence is

$$(14 \text{ sec} + 32 \text{ sec}) \left(\frac{1 \text{ min}}{60 \text{ sec}} \right) = 0.77 \text{ min}$$

The time required to fill one truck is

$$(69 \text{ residences})(0.77 \text{ min}) = 53 \text{ min}$$

One truck and crew can service

$$\left(480 \frac{\text{min}}{\text{day}} \right) \left(\frac{69 \text{ residences}}{53 \text{ min}} \right)$$
$$= 625 \text{ residences/day} \quad (630 \text{ residences/day})$$

SOLUTION 14

The landfill volume is

$$(0.5) \left(\begin{array}{c} (1200 \text{ ft} - 80 \text{ ft} - 80 \text{ ft}) \\ \times (1600 \text{ ft} - 80 \text{ ft} - 80 \text{ ft}) \\ + (1200 \text{ ft})(1600 \text{ ft}) \end{array} \right) (80 \text{ ft})$$
$$= 1.4 \times 10^8 \text{ ft}^3$$

The annual waste mass landfilled is

$$(0.75)(215{,}000 \text{ people}) \left(4.6 \frac{\text{lbm}}{\text{capita day}} \right) \left(365 \frac{\text{day}}{\text{yr}} \right)$$
$$= 2.7 \times 10^8 \text{ lbm/yr}$$

The annual in-place waste volume landfilled is

$$\frac{2.7 \times 10^8 \frac{\text{lbm}}{\text{yr}}}{50 \frac{\text{lbm}}{\text{ft}^3}} = 5.4 \times 10^6 \text{ ft}^3/\text{yr}$$

The annual cover volume is

$$\frac{5.4 \times 10^6 \ \frac{ft^3}{yr}}{4.5} = 1.2 \times 10^6 \ ft^3/yr$$

The annual landfill total volume is

$$5.4 \times 10^6 \ \frac{ft^3}{yr} + 1.2 \times 10^6 \ \frac{ft^3}{yr} = 6.6 \times 10^6 \ ft^3/yr$$

The landfill operating life is

$$\frac{1.4 \times 10^8 \ ft^3}{6.6 \times 10^6 \ \frac{ft^3}{yr}} = 21 \ yr$$

The answer is (C).

Why Other Options Are Wrong

(A) This incorrect solution reverses the cover-to-fill ratio. Definitions and equations are unchanged from the correct solution.

The landfill volume is

$$(0.5)\begin{pmatrix}(1200 \ ft - 80 \ ft - 80 \ ft) \\ \times (1600 \ ft - 80 \ ft - 80 \ ft) \\ + (1200 \ ft)(1600 \ ft)\end{pmatrix}(80 \ ft)$$
$$= 1.4 \times 10^8 \ ft^3$$

The annual waste mass landfilled is

$$(0.75)(215{,}000 \ \text{people})\left(4.6 \ \frac{lbm}{\text{capita day}}\right)\left(365 \ \frac{day}{yr}\right)$$
$$= 2.7 \times 10^8 \ lbm/yr$$

The annual in-place waste volume landfilled is

$$\frac{2.7 \times 10^8 \ \frac{lbm}{yr}}{50 \ \frac{lbm}{ft^3}} = 5.4 \times 10^6 \ ft^3/yr$$

The annual cover volume is

$$\left(5.4 \times 10^6 \ \frac{ft^3}{yr}\right)(4.5) = 2.4 \times 10^7 \ ft^3/yr$$

The annual landfill total volume is

$$5.4 \times 10^6 \ \frac{ft^3}{yr} + 2.4 \times 10^7 \ \frac{ft^3}{yr} = 2.9 \times 10^7 \ ft^3/yr$$

The landfill operating life is

$$\frac{1.4 \times 10^8 \ ft^3}{2.9 \times 10^7 \ \frac{ft^3}{yr}} = 5 \ yr$$

(B) This solution is incorrect because both the recycled-to-landfilled ratio and the cover-to-fill ratio are reversed. The landfilled mass is based on the percent recycled instead of the percent landfilled. Definitions and equations are unchanged from the correct solution.

The landfill volume is

$$(0.5)\begin{pmatrix}(1200 \ ft - 80 \ ft - 80 \ ft) \\ \times (1600 \ ft - 80 \ ft - 80 \ ft) \\ + (1200 \ ft)(1600 \ ft)\end{pmatrix}(80 \ ft)$$
$$= 1.4 \times 10^8 \ ft^3$$

The annual waste mass landfilled is

$$(0.25)(215{,}000 \ \text{people})\left(4.6 \ \frac{lbm}{\text{capita day}}\right)\left(365 \ \frac{day}{yr}\right)$$
$$= 9.0 \times 10^7 \ lbm/yr$$

The annual in-place waste volume landfilled is

$$\frac{9.0 \times 10^7 \ \frac{lbm}{yr}}{50 \ \frac{lbm}{ft^3}} = 1.8 \times 10^6 \ ft^3/yr$$

The annual cover volume is

$$\left(1.8 \times 10^6 \ \frac{ft^3}{yr}\right)(4.5) = 8.1 \times 10^6 \ ft^3/yr$$

The annual landfill total volume is

$$1.8 \times 10^6 \ \frac{ft^3}{yr} + 8.1 \times 10^6 \ \frac{ft^3}{yr} = 9.9 \times 10^6 \ ft^3/yr$$

The landfill operating life is

$$\frac{1.4 \times 10^8 \ ft^3}{9.9 \times 10^6 \ \frac{ft^3}{yr}} = 14 \ yr$$

(D) This choice is incorrect because the annual waste mass landfilled is calculated based on percent recycled. Definitions and equations are unchanged from the correct solution.

The landfill volume is

$$(0.5)\begin{pmatrix}(1200 \ ft - 80 \ ft - 80 \ ft) \\ \times (1600 \ ft - 80 \ ft - 80 \ ft) \\ + (1200 \ ft)(1600 \ ft)\end{pmatrix}(80 \ ft)$$
$$= 1.4 \times 10^8 \ ft^3$$

The annual waste mass landfilled is

$$(0.25)(215{,}000 \ \text{people})\left(4.6 \ \frac{lbm}{\text{capita day}}\right)\left(365 \ \frac{day}{yr}\right)$$
$$= 9.0 \times 10^7 \ lbm/yr$$

The annual in-place waste volume landfilled is

$$\frac{9.0 \times 10^7 \; \frac{\text{lbm}}{\text{yr}}}{50 \; \frac{\text{lbm}}{\text{ft}^3}} = 1.8 \times 10^6 \; \text{ft}^3/\text{yr}$$

The annual cover volume is

$$\frac{1.8 \times 10^6 \; \frac{\text{ft}^3}{\text{yr}}}{4.5} = 4.0 \times 10^5 \; \text{ft}^3/\text{yr}$$

The annual landfill total volume is

$$1.8 \times 10^6 \; \frac{\text{ft}^3}{\text{yr}} + 4.0 \times 10^5 \; \frac{\text{ft}^3}{\text{yr}} = 2.2 \times 10^6 \; \text{ft}^3/\text{yr}$$

The landfill operating life is

$$\frac{1.4 \times 10^8 \; \text{ft}^3}{2.2 \times 10^6 \; \text{ft}^3/\text{yr}} = 64 \; \text{yr}$$

SOLUTION 15

K_H	Henry's constant	atm
K'_H	Henry's constant	–
MW	molecular weight	g/mol
R^*	universal gas constant	0.082 atm·L/mol·K
T	temperature	K
ρ_w	water density	1000 g/L

The Henry's constant for methylene chloride is 177 atm.

$$K'_H = \frac{K_H \text{MW}_w}{\rho_{\text{water}} R^* T}$$

$$= \frac{(177 \; \text{atm}) \left(18 \; \frac{\text{g}}{\text{mol}}\right)}{\left(1000 \; \frac{\text{g}}{\text{L}}\right) \left(0.082 \; \frac{\text{atm·L}}{\text{mol·K}}\right) (25°\text{C} + 273°)}$$

$$= 0.13 \; \text{unitless}$$

S	stripping factor	–
V_a/V_w	air-to-water ratio	–

$$\frac{V_a}{V_w} = \frac{S}{K'_H} = \frac{3.5}{0.13}$$

$$= 27$$

Q	water flow rate	gal/min
Q_a	air flow rate	ft³/min

$$Q_a = Q\left(\frac{V_a}{V_w}\right)$$

$$= \left(135 \; \frac{\text{gal}}{\text{min}}\right) \left(0.134 \; \frac{\text{ft}^3}{\text{gal}}\right) (27)$$

$$= 488 \; \text{ft}^3/\text{min} \quad (490 \; \text{ft}^3/\text{min})$$

The answer is (C).

Why Other Options Are Wrong

(A) This incorrect choice includes the concentration ratio in the calculation of air-to-water ratio. Other assumptions, definitions, and equations are unchanged from the correct solution.

The Henry's constant for methylene chloride is 177 atm.

$$K'_H = \frac{(177 \; \text{atm}) \left(18 \; \frac{\text{g}}{\text{mol}}\right)}{\left(1000 \; \frac{\text{g}}{\text{L}}\right) \left(0.082 \; \frac{\text{atm·L}}{\text{mol·K}}\right) (25°\text{C} + 273°)}$$

$$= 0.13$$

C	effluent concentration	mg/L
C_o	influent concentration	mg/L

$$\frac{V_a}{V_w} = \frac{SC}{K'_H C_o} = \frac{(3.5) \left(0.100 \; \frac{\text{mg}}{\text{L}}\right)}{(0.13) \left(12 \; \frac{\text{mg}}{\text{L}}\right)}$$

$$= 0.22$$

$$Q_a = Q\left(\frac{V_a}{V_w}\right)$$

$$= \left(135 \; \frac{\text{gal}}{\text{min}}\right) \left(0.134 \; \frac{\text{ft}^3}{\text{gal}}\right) (0.22)$$

$$= 4.0 \; \text{ft}^3/\text{min}$$

(B) This incorrect choice uses the stripping factor as the air-to-water ratio. Other definitions are unchanged from the correct solution.

$$Q_a = QS$$

$$= \left(135 \; \frac{\text{gal}}{\text{min}}\right) (3.5) \left(0.134 \; \frac{\text{ft}^3}{\text{gal}}\right)$$

$$= 63 \; \text{ft}^3/\text{min}$$

(D) This incorrect choice uses Henry's constant with units of atm instead of unitless and inverts the air-to-water ratio. Other assumptions, definitions, and equations are the same as used for the correct solution.

The Henry's constant for methylene chloride is 177 atm.

$$\frac{V_w}{V_a} = \frac{S}{K_H} = \frac{3.5}{177 \; \text{atm}}$$

$$= 0.020$$

Units are ignored.

$$Q_a = Q\left(\frac{V_a}{V_w}\right)$$

$$= \frac{\left(135 \; \frac{\text{gal}}{\text{min}}\right) \left(0.134 \; \frac{\text{ft}^3}{\text{gal}}\right)}{0.020}$$

$$= 905 \; \text{ft}^3/\text{min} \quad (900 \; \text{ft}^3/\text{min})$$

SOLUTION 16

$\dot{m}_{in}$	feed mass flow rate	kg/h
$\dot{m}_{out}$	POHC mass flow rate from the stack	kg/h

$$\text{DRE} = \left(\frac{W_{in} - W_{out}}{W_{in}}\right) \times 100\%$$

$$= \left(\frac{4.12 \frac{\text{kg}}{\text{h}} - 0.000\,74 \frac{\text{kg}}{\text{h}}}{4.12 \frac{\text{kg}}{\text{h}}}\right) \times 100\%$$

$$= 99.98\%$$

The answer is (D).

Why Other Options Are Wrong

(A) This incorrect solution uses the difference between the mass flow rate to the pollution control equipment and the mass flow rate from the stack. The difference is divided by the feed mass flow rate. Other assumptions, definitions, and equations are unchanged from the correct solution.

$\dot{m}_{in}$	POHC mass flow rate to the stack	kg/h

$$\text{DRE} = \left(\frac{0.43 \frac{\text{kg}}{\text{h}} - 0.000\,74 \frac{\text{kg}}{\text{h}}}{4.12 \frac{\text{kg}}{\text{h}}}\right) \times 100\%$$

$$= 10.42\%$$

(B) This incorrect solution uses the difference between the feed rate and mass flow rate from the incinerator instead of from the stack. This calculates the destruction efficiency. Other assumptions, definitions, and equations are unchanged from correct solution.

$\dot{m}_{out}$	POHC mass flow rate to the stack	kg/h

$$\text{DRE} = \left(\frac{4.12 \frac{\text{kg}}{\text{h}} - 0.43 \frac{\text{kg}}{\text{h}}}{4.12 \frac{\text{kg}}{\text{h}}}\right) \times 100\%$$

$$= 89.56\%$$

(C) This incorrect solution uses the difference between the mass flow rate to the pollution control equipment and from the stack. This calculates the removal efficiency. Other assumptions, definitions, and equations are unchanged from the correct solution.

$\dot{m}_{in}$	POHC mass flow rate to stack	kg/h
$\dot{m}_{out}$	POHC mass flow rate from stack	kg/h

$$\text{DRE} = \left(\frac{0.43 \frac{\text{kg}}{\text{h}} - 0.000\,74 \frac{\text{kg}}{\text{h}}}{0.43 \frac{\text{kg}}{\text{h}}}\right) \times 100\%$$

$$= 99.83\%$$

GROUNDWATER AND WELL FIELDS

SOLUTION 17

g	gravitational constant	9.81 m/s^2
k_i	intrinsic permeability	cm^2
K_w	hydraulic conductivity for water	cm/s
μ_w	water viscosity	0.013 07 g/cm·s at 10°C
ρ_w	water density	0.9997 g/cm^3 at 10°C

$$k_i = \frac{K_w \mu_w}{\rho_w g}$$

$$= \frac{\left(2.0 \times 10^{-4} \frac{\text{cm}}{\text{s}}\right)\left(0.013\,07 \frac{\text{g}}{\text{cm·s}}\right)}{\left(0.9997 \frac{\text{g}}{\text{cm}^3}\right)\left(9.81 \frac{\text{m}}{\text{s}^2}\right)\left(100 \frac{\text{cm}}{\text{m}}\right)}$$

$$= 2.67 \times 10^{-9} \text{ cm}^2$$

K_{NAPL}	hydraulic conductivity of NAPL	cm/s
μ_f	NAPL viscosity	0.066 g/cm·s at 10°C
ρ_f	NAPL density	0.92 g/cm^3 at 10°C

$$K_{\text{NAPL}} = \frac{k_i g \rho_f}{\mu_f}$$

$$= \frac{(2.67 \times 10^{-9} \text{ cm}^2)\left(9.81 \frac{\text{m}}{\text{s}^2}\right) \times \left(0.92 \frac{\text{g}}{\text{cm}^3}\right)\left(100 \frac{\text{cm}}{\text{m}}\right)}{0.066 \frac{\text{g}}{\text{cm·s}}}$$

$$= 3.6 \times 10^{-5} \text{ cm/s}$$

The answer is (B).

Why Other Options Are Wrong

(A) This incorrect solution inverts NAPL viscosity and density. Other assumptions, definitions, and equations are unchanged from the correct solution.

$$k_i = \frac{\left(2.0 \times 10^{-4} \frac{\text{cm}}{\text{s}}\right)\left(0.013\,07 \frac{\text{g}}{\text{cm·s}}\right)}{\left(0.9997 \frac{\text{g}}{\text{cm}^3}\right)\left(9.81 \frac{\text{m}}{\text{s}^2}\right)\left(100 \frac{\text{cm}}{\text{m}}\right)}$$

$$= 2.67 \times 10^{-9} \text{ cm}^2$$

$$K_{NAPL} = \frac{k_i g \mu_{NAPL}}{\rho_{NAPL}}$$

$$= \frac{(2.67 \times 10^{-9} \text{ cm}^2)\left(9.81 \frac{\text{m}}{\text{s}^2}\right)}{0.92 \frac{\text{g}}{\text{cm}^3}} \times \left(0.066 \frac{\text{g}}{\text{cm}\cdot\text{s}}\right)\left(100 \frac{\text{cm}}{\text{m}}\right)$$

$$= 1.9 \times 10^{-7} \text{ cm/s}$$

Note that units do not work.

(C) This incorrect solution uses density of water instead of NAPL. Other assumptions, definitions, and equations are unchanged from the correct solution.

$$k_i = \frac{\left(2.0 \times 10^{-4} \frac{\text{cm}}{\text{s}}\right)\left(0.01307 \frac{\text{g}}{\text{cm}\cdot\text{s}}\right)}{\left(0.9997 \frac{\text{g}}{\text{cm}^3}\right)\left(9.81 \frac{\text{m}}{\text{s}^2}\right)\left(100 \frac{\text{cm}}{\text{m}}\right)}$$

$$= 2.67 \times 10^{-9} \text{ cm}^2$$

$$K_{NAPL} = \frac{k_i g \rho_w}{\mu_{NAPL}}$$

$$= \frac{(2.67 \times 10^{-9} \text{ cm}^2)\left(9.81 \frac{\text{m}}{\text{s}^2}\right)}{0.066 \frac{\text{g}}{\text{cm}\cdot\text{s}}} \times \left(0.9997 \frac{\text{g}}{\text{cm}^3}\right)\left(100 \frac{\text{cm}}{\text{m}}\right)$$

$$= 4.0 \times 10^{-5} \text{ cm/s}$$

(D) This incorrect solution uses dynamic viscosity of water instead of NAPL. Other assumptions, definitions, and equations are unchanged from the correct solution.

$$k_i = \frac{\left(2.0 \times 10^{-4} \frac{\text{cm}}{\text{s}}\right)\left(0.01307 \frac{\text{g}}{\text{cm}\cdot\text{s}}\right)}{\left(0.9997 \frac{\text{g}}{\text{cm}^3}\right)\left(9.81 \frac{\text{m}}{\text{s}^2}\right)\left(100 \frac{\text{cm}}{\text{m}}\right)}$$

$$= 2.67 \times 10^{-9} \text{ cm}^2$$

$$K_{NAPL} = \frac{k_i g \rho_{NAPL}}{\mu_w}$$

$$= \frac{(2.67 \times 10^{-9} \text{ cm}^2)\left(9.81 \frac{\text{m}}{\text{s}^2}\right)}{0.01307 \frac{\text{g}}{\text{cm}\cdot\text{s}}} \times \left(0.92 \frac{\text{g}}{\text{cm}^3}\right)\left(100 \frac{\text{cm}}{\text{m}}\right)$$

$$= 1.8 \times 10^{-4} \text{ cm/s}$$

SOLUTION 18

h_o aquifer thickness   ft
h h_o − observation well drawdown   ft

$$h = 18 \text{ ft} - (14 \text{ in})\left(\frac{1 \text{ ft}}{12 \text{ in}}\right) = 16.83 \text{ ft}$$

K hydraulic conductivity   ft/day
Q pumping rate   gal/min
r distance from pumped well to observation well   30 ft
r_o radius of influence   ft

$$Q = \frac{K\pi(h_o^2 - h^2)}{\ln \frac{r_o}{r}}$$

$$= \frac{\left(20 \frac{\text{gal}}{\text{min}}\right)\left(0.134 \frac{\text{ft}^3}{\text{gal}}\right)\left(1440 \frac{\text{min}}{\text{day}}\right)}{\left(7.2 \frac{\text{ft}}{\text{day}}\right)\pi\left((18 \text{ ft})^2 - (16.83 \text{ ft})^2\right)} = \ln\frac{r_o}{30 \text{ ft}}$$

$r_o = 38 \text{ ft}$

The answer is (C).

Why Other Options Are Wrong

(A) This incorrect solution violates logarithm laws by distributing the natural log to each term inside the parentheses. Other assumptions, definitions, and equations are the same as those used in the correct solution.

$$h = 18 \text{ ft} - (14 \text{ in})\left(\frac{1 \text{ ft}}{12 \text{ in}}\right) = 16.83 \text{ ft}$$

$$\frac{\left(20 \frac{\text{gal}}{\text{min}}\right)\left(0.134 \frac{\text{ft}^3}{\text{gal}}\right)\left(1440 \frac{\text{min}}{\text{day}}\right)}{\left(7.2 \frac{\text{ft}}{\text{day}}\right)\pi\left((18 \text{ ft})^2 - (16.83 \text{ ft})^2\right)} = \ln\frac{r_o}{30 \text{ ft}}$$

$$\ln r_o = \frac{(\ln 30 \text{ ft})\left(7.2 \frac{\text{ft}}{\text{day}}\right)\pi\left((18 \text{ ft})^2 - (16.83 \text{ ft})^2\right)}{\left(20 \frac{\text{gal}}{\text{min}}\right)\left(0.134 \frac{\text{ft}^3}{\text{gal}}\right)\left(1440 \frac{\text{min}}{\text{day}}\right)}$$

$r_o = 2.3 \text{ ft}$

(B) This solution is incorrect because the unit conversion from gal to ft^3 is not made for flow. Other assumptions, definitions, and equations are the same as those used in the correct solution.

$$h = 18 \text{ ft} - (14 \text{ in})\left(\frac{1 \text{ ft}}{12 \text{ in}}\right) = 16.83 \text{ ft}$$

$$\frac{\left(20 \frac{\text{gal}}{\text{min}}\right)\left(1440 \frac{\text{min}}{\text{day}}\right)}{\left(7.2 \frac{\text{ft}}{\text{day}}\right)\pi\left((18 \text{ ft})^2 - (16.83 \text{ ft})^2\right)} = \ln\frac{r_o}{30 \text{ ft}}$$

$$r_o = 31 \text{ ft}$$

Note that units are ignored.

(D) This solution is incorrect because the pumping time is included with the flow. Other assumptions, definitions, and equations are the same as those used in the correct solution.

$$h = 18 \text{ ft} - (14 \text{ in})\left(\frac{1 \text{ ft}}{12 \text{ in}}\right) = 16.83 \text{ ft}$$

$$= \frac{\left(20 \frac{\text{gal}}{\text{min}}\right)(10 \text{ hr})\left(60 \frac{\text{min}}{\text{hr}}\right)\left(0.134 \frac{\text{ft}^3}{\text{gal}}\right)}{\left(7.2 \frac{\text{ft}}{\text{day}}\right)\pi\left((18 \text{ ft})^2 - (16.83 \text{ ft})^2\right)}$$
$$\ln \frac{r_o}{30 \text{ ft}}$$

$$r_o = 53 \text{ ft}$$

Note that time units do not cancel and are ignored.

SOLUTION 19

d_i thickness of layer i cm
K hydraulic conductivity   cm/s

The minimum typical hydraulic conductivity for each layer is

layer	K (cm/s)
1	5×10^{-4}
2	5×10^{-8}
3	2.5×10^{-5}
4	2.5×10^{-7}

$$K_{\text{overall}} = \frac{d_1 + d_2 + \cdots d_n}{\frac{d_1}{K_1} + \frac{d_2}{K_2} + \cdots \frac{d_n}{K_n}}$$

layer	d (cm)	d/K (s)
1	70	1.40×10^5
2	109	2.18×10^9
3	88	3.52×10^6
4	46	1.84×10^8

$$K_{\text{overall}} = \frac{70 \text{ cm} + 109 \text{ cm} + 88 \text{ cm} + 46 \text{ cm}}{1.40 \times 10^5 \text{ s} + 2.18 \times 10^9 \text{ s} + 3.52 \times 10^6 \text{ s} + 1.84 \times 10^8 \text{ s}}$$
$$= 1.3 \times 10^{-7} \text{ cm/s}$$

The answer is (B).

Why Other Options Are Wrong

(A) This incorrect solution assumes that the overall hydraulic conductivity is equal to that of the least permeable layer. Definitions are unchanged from the correct solution.

$$K_{\text{overall}} = 5.0 \times 10^{-8} \text{ cm/s}$$

(C) This incorrect calculation is based on the simple average of the typical hydraulic conductivities. Definitions are unchanged from the correct solution.

The minimum typical hydraulic conductivity for each layer is

layer	K (cm/s)
1	5×10^{-4}
2	5×10^{-8}
3	2.5×10^{-5}
4	2.5×10^{-7}

$$K_{\text{overall}} = \frac{5 \times 10^{-4} \frac{\text{cm}}{\text{s}} + 5 \times 10^{-8} \frac{\text{cm}}{\text{s}} + 2.5 \times 10^{-5} \frac{\text{cm}}{\text{s}} + 2.5 \times 10^{-7} \frac{\text{cm}}{\text{s}}}{4}$$
$$= 1.3 \times 10^{-4} \text{ cm/s}$$

(D) This incorrect solution assumes that the overall hydraulic conductivity is equal to that of the most permeable layer. Definitions are unchanged from the correct solution.

$$K_{\text{overall}} = 5.0 \times 10^{-4} \text{ cm/s}$$

SOLUTION 20

The groundwater elevation contour lines are drawn to make it possible to determine the groundwater gradient.

i groundwater gradient   –
ΔL distance between groundwater contour lines of interest   ft
Δh elevation change over distance L ft

$$i = \frac{\Delta h}{\Delta L}$$
$$= \frac{3210 \text{ ft} - 3207 \text{ ft}}{2300 \text{ ft}}$$
$$= 0.0013$$

K hydraulic conductivity   ft/day
n_e effective porosity   –
r_f retardation factor   –
v_s solute velocity   ft/day

$$v_s = \frac{Ki}{n_e r_f}$$
$$= \frac{\left(0.83 \frac{\text{ft}}{\text{day}}\right)(0.0013)}{(0.37)(1.94)}$$
$$= 0.0015 \text{ ft/day}$$

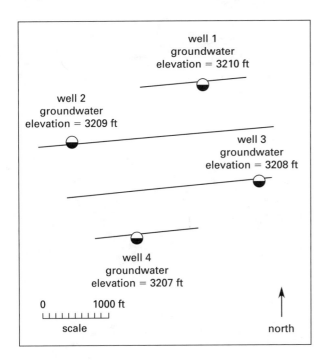

The answer is (B).

Why Other Options Are Wrong

(A) This incorrect solution does not include porosity. This calculates the Darcy velocity, not actual velocity. The figure and other assumptions, definitions, and equations are unchanged from the correct solution.

$$i = \frac{3210 \text{ ft} - 3207 \text{ ft}}{2300 \text{ ft}} = 0.0013$$

$$v_s = \frac{Ki}{r_f} = \frac{\left(0.83 \ \frac{\text{ft}}{\text{day}}\right)(0.0013)}{1.94}$$

$$= 0.00056 \text{ ft/day}$$

(C) This incorrect solution multiplies by, instead of divides by, the retardation factor and does not include porosity. The figure and other assumptions, definitions, and equations are unchanged from the correct solution.

$$i = \frac{3210 \text{ ft} - 3207 \text{ ft}}{2300 \text{ ft}} = 0.0013$$

$$v_s = Kir_f = \left(0.83 \ \frac{\text{ft}}{\text{day}}\right)(0.0013)(1.94)$$

$$= 0.0021 \text{ ft/day}$$

(D) This incorrect solution does not include the retardation factor. The figure and other assumptions, definitions, and equations are unchanged from the correct solution.

$$i = \frac{3210 \text{ ft} - 3207 \text{ ft}}{2300 \text{ ft}} = 0.0013$$

$$v_s = \frac{Ki}{n_e} = \frac{\left(0.83 \ \frac{\text{ft}}{\text{day}}\right)(0.0013)}{0.37}$$

$$= 0.0029 \text{ ft/day}$$

Depth Solutions

WASTEWATER TREATMENT

SOLUTION 21

The pretreatment system's capacity should be able to meet the demand 90% of the time.

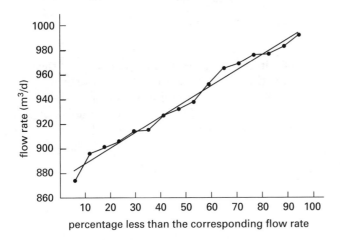

From the illustration, the desired flow rate = 990 m³/d.

The answer is (C).

Why Other Options Are Wrong

(A) This incorrect choice uses 90% of the average value of the average weekly flows.

week	weekly average (m³/d)	week	weekly average (m³/d)
1	976	9	906
2	901	10	927
3	938	11	969
4	965	12	932
5	977	13	983
6	992	14	896
7	874	15	915
8	914	16	952
			15 017

Desired flow rate is

$$\frac{\left(15\,017\ \frac{\text{m}^3}{\text{d}}\right)(0.90)}{16} = 845\ \text{m}^3/\text{d} \quad (850\ \text{m}^3/\text{d})$$

(B) This incorrect choice uses the flow rate from the illustration at 10%.

From the illustration, the desired flow rate = 885 m³/d (890 m³/d).

(D) This incorrect choice divides the average of the average weekly flows by 90%.

week	weekly average (m³/d)	week	weekly average (m³/d)
1	976	9	906
2	901	10	927
3	938	11	969
4	965	12	932
5	977	13	983
6	992	14	896
7	874	15	915
8	914	16	952
			15 017

Desired flow rate is

$$\frac{15\,017\ \frac{\text{m}^3}{\text{d}}}{(16)(0.90)} = 1043\ \text{m}^3/\text{d} \quad (1100\ \text{m}^3/\text{d})$$

SOLUTION 22

Assume that the typical person generates 0.2 lbm BOD/day-person.

$$\text{PE} = \frac{\text{BOD loading rate}}{\text{per capita BOD generation rate}}$$

The BOD loading rate is

$$\left(0.2\ \frac{\text{lbm BOD}}{\text{day person}}\right)\left(4\ \frac{\text{PE}}{\text{home}}\right)(1750\ \text{homes})$$
$$= 1400\ \text{lbm BOD/day}$$

The answer is (B).

Why Other Options Are Wrong

(A) This incorrect solution divides by population equivalents instead of multiplying. The units did not work, but "homes" and "people/home" are atypical units that some responders may incorrectly feel justified in ignoring. Other assumptions are unchanged from the correct solution.

The BOD loading rate is

$$\frac{(1750 \text{ homes})\left(0.2 \frac{\text{lbm BOD}}{\text{day person}}\right)}{4 \text{ PE}} = 88 \text{ lbm BOD/day}$$

(C) This incorrect solution confuses the units of the typical per capita BOD generation rate and ignores the units for PE and "homes." Other assumptions are unchanged from the correct solution.

The BOD loading rate is

$$\frac{1750 \text{ homes}}{(4 \text{ PE})\left(0.2 \frac{\text{people lbm BOD}}{\text{day}}\right)} = 2188 \text{ lbm BOD/day} \quad (2200 \text{ lbm BOD/day})$$

(D) This incorrect solution confuses the units of the typical per capita BOD generation rate. Other assumptions are unchanged from the correct solution.

The BOD loading rate is

$$\frac{\left(4 \frac{\text{people}}{\text{home}}\right)(1750 \text{ homes})}{0.2 \frac{\text{people lbm BOD}}{\text{day}}} = 35{,}000 \text{ lbm BOD/day}$$

SOLUTION 23

When the initial pH is less than 8.3, the bicarbonate alkalinity and total alkalinity are equal. 1 mL of 0.03 N H_2SO_4 will titrate 1.5 mg alkalinity as $CaCO_3$.

M	bicarbonate alkalinity	mg/L as $CaCO_3$
M_{total}	total alkalinity	mg/L as $CaCO_3$
V_{sample}	sample volume	mL
$V_{titrant}$	titrant volume	mL
W	acid equivalent as $CaCO_3$	mg as $CaCO_3$/ mL acid

$$M = M_{total} = \frac{W V_{titrant}}{V_{sample}}$$

$$= \frac{(1.5 \text{ mg alk as } CaCO_3)(14.5 \text{ mL } 0.03 \text{ N } H_2SO_4) \times \left(1000 \frac{\text{mL}}{\text{L}}\right)}{(1 \text{ mL } 0.03 \text{ N } H_2SO_4)(500 \text{ mL sample})}$$

$$= 44 \text{ mg/L as } CaCO_3$$

The answer is (C).

Why Other Options Are Wrong

(A) This incorrect choice calculates alkalinity correctly, but then attempts to correct it for a 500 mL sample. The concentration is independent of the sample size. Other assumptions, definitions, and equations are unchanged from the correct solution.

$$M = \frac{(1.5 \text{ mg alk as } CaCO_3) \times (14.5 \text{ mL } 0.03 \text{ NH}_2SO_4)\left(1000 \frac{\text{mL}}{\text{L}}\right)}{(1 \text{ mL } 0.03 \text{ NH}_2SO_4)(500 \text{ mL sample})}$$

$$= 44 \text{ mg/L as } CaCO_3$$

For a 500 mL sample, the bicarbonate alkalinity is

$$(0.5)\left(44 \frac{\text{mg}}{\text{L}} \text{ as } CaCO_3\right) = 22 \text{ mg/L as } CaCO_3$$

(B) This incorrect choice assumes that 1 mL of the 0.03 N H_2SO_4 titrant would neutralize 1 mg of alkalinity as $CaCO_3$. This is only true if the standard titrant concentration of 0.02 N H_2SO_4 is used. Other assumptions, definitions, and equations are unchanged from the correct solution.

Assume that 1 ml of 0.03 N H_2SO_4 will titrate 1.0 mg alkalinity as $CaCO_3$.

$$M = \frac{(1.0 \text{ mg alk as } CaCO_3) \times (14.5 \text{ mL } 0.03 \text{ NH}_2SO_4)\left(1000 \frac{\text{mL}}{\text{L}}\right)}{(1 \text{ mL } 0.03 \text{ NH}_2SO_4)(500 \text{ mL sample})}$$

$$= 29 \text{ mg/L as } CaCO_3$$

(D) This incorrect solution calculates alkalinity in mg/L as HCO_3^- and then converts units to mg/L as $CaCO_3$. The titrant used gives alkalinity units in mg as $CaCO_3$, and no subsequent conversion is required. Other assumptions and definitions are unchanged from the correct solution.

$$HCO_3^- = \frac{(1.5 \text{ mg } HCO_3^-)(14.5 \text{ mL } 0.03 \text{ N } H_2SO_4) \times \left(1000 \frac{\text{mL}}{\text{L}}\right)}{(1 \text{ mL } 0.03 \text{ N } H_2SO_4)(500 \text{ mL sample})}$$

$$= 44 \text{ mg/L}$$

The mole weight of HCO_3^- is 61 mg/mmol.

The mole weight of $CaCO_3$ is 100 mg/mmol.

$$M = \frac{\left(44 \frac{\text{mg } HCO_3^-}{\text{L}}\right)\left(100 \frac{\text{mg } CaCO_3}{\text{mmol}}\right)}{61 \frac{\text{mg } HCO_3^-}{\text{mmol}}}$$

$$= 72 \text{ mg/L as } CaCO_3$$

SOLUTION 24

Because decreasing temperature results in a longer reaction time, take 21°C as the reference temperature (T_1) and 17°C as the temperature of interest (T_2).

t	time	min
T	temperature	K
E	activation energy	6400 cal/mol for aqueous chlorine at pH 8.5
R	gas law constant	1.99 cal/mol·K

$$T_1 = (21°C + 273°) = 294K$$
$$T_2 = (17°C + 273°) = 290K$$

$$\ln\frac{t_1}{t_2} = \frac{E(T_2 - T_1)}{RT_1T_2}$$

$$= \frac{\left(6400 \ \frac{\text{cal}}{\text{mol}}\right)(290K - 294K)}{\left(1.99 \ \frac{\text{cal}}{\text{mol·K}}\right)(294K)(290K)}$$

$$= -0.15$$

$$t_2 = \frac{t_1}{e^{-0.15}} = \frac{23 \text{ min}}{e^{-0.15}} = 27 \text{ min}$$

The answer is (C).

Why Other Options Are Wrong

(A) This incorrect choice uses °C for temperature instead of K, takes the natural log instead of raising e to the whole-number power, and reverses the temperature values in the equation. Other assumptions, definitions, and equations are the same as for the correct solution.

$$T_1 = 17°C$$
$$T_2 = 21°C$$

$$\ln\frac{t_1}{t_2} = \frac{\left(6400 \ \frac{\text{cal}}{\text{mol}}\right)(21°C - 17°C)}{\left(1.99 \ \frac{\text{cal}}{\text{mol·K}}\right)(21°C)(17°C)}$$

$$= 36$$

Note that the temperature units do not cancel.

$$t_2 = \frac{t_1}{\ln 36} = \frac{23 \text{ min}}{\ln 36}$$
$$= 6.4 \text{ min}$$

(B) This incorrect choice reverses the temperature values in the equation. Other assumptions, definitions, and equations are unchanged from the correct solution.

$$T_1 = 17°C + 273° = 290K$$
$$T_2 = 21°C + 273° = 294K$$

$$\ln\frac{t_1}{t_2} = \frac{\left(6400 \ \frac{\text{cal}}{\text{mol}}\right)(294K - 290K)}{\left(1.99 \ \frac{\text{cal}}{\text{mol·K}}\right)(290K)(294K)}$$

$$= 0.15$$

$$t_2 = \frac{23 \text{ min}}{e^{0.15}}$$
$$= 20 \text{ min}$$

(D) This incorrect choice uses activation energy for chloramines at pH 8.5 instead of for aqueous chlorine. Other assumptions, definitions, and equations are unchanged from the correct solution.

E	activation energy	14 000 cal/mol for chloramines at pH 8.5

$$T_1 = 21°C + 273° = 294K$$
$$T_2 = 17°C + 273° = 290K$$

$$\ln\frac{t_1}{t_2} = \frac{\left(14\,000 \ \frac{\text{cal}}{\text{mol}}\right)(290K - 294K)}{\left(1.99 \ \frac{\text{cal}}{\text{mol·K}}\right)(294K)(290K)}$$

$$= -0.33$$

$$t_2 = \frac{23 \text{ min}}{e^{-0.33}} = 32 \text{ min}$$

SOLUTION 25

The 1977 amendments to the Clean Water Act (CWA) included a list of 65 priority pollutants (specific chemicals and classes of chemicals) to be used for defining toxic substances. The original list has been expanded to the current list of 129 priority pollutants. As defined by the CWA, priority pollutants are those chemicals with relatively high toxicity and high production volume. Although they may not be the most toxic chemicals, because of their generally widespread use and relative toxicity they have received special attention by regulators.

The answer is (B).

Why Other Options Are Wrong

(A) Two criteria must be satisfied to establish a chemical or group of chemicals as priority pollutants. One of these is toxicity, but if the chemical is produced in small quantities, its potential hazard is minimized. Both toxicity and production volume are necessary for a chemical to present a significant potential risk. Consequently, priority pollutants do not include many of the most toxic of known chemicals.

(C) Specific criteria of toxicity, flammability, corrosivity, or reactivity are associated with hazardous waste under RCRA. These criteria are not commonly used to define priority pollutants.

(D) No direct regulatory relationship exists between priority pollutants and the National Priorities List (NPL). Priority pollutants may be found at NPL sites, but this does not define them as priority pollutants.

SOLUTION 26

d_p	mean particle diameter	mm
f	Darcy friction factor	–
g	gravitational constant	9.81 m/s²
k	Camp constant	–
SG_p	grit specific gravity	–
v	horizontal velocity	m/s

$$v = \sqrt{\frac{8kgd_p(SG_p - 1)}{f}}$$

$$= \sqrt{\frac{(8)(0.05)\left(9.81 \frac{m}{s^2}\right)(0.22 \text{ mm}) \times \left(\frac{1 \text{ m}}{1000 \text{ mm}}\right)(2.65 - 1)}{0.03}}$$

$$= 0.22 \text{ m/s}$$

A	channel cross sectional area	ft²
d	depth	ft
Q	flow rate	gal/day
w	width	ft

$$Q = Av = wdv$$

$$w = \frac{Q}{dv}$$

$$= \frac{\left(3.5 \times 10^6 \frac{\text{gal}}{\text{day}}\right)\left(0.134 \frac{\text{ft}^3}{\text{gal}}\right)}{(4 \text{ ft})\left(0.22 \frac{m}{s}\right)\left(3.28 \frac{\text{ft}}{m}\right)\left(86{,}400 \frac{\text{sec}}{\text{day}}\right)}$$

$$= 1.9 \text{ ft}$$

The answer is (B).

Why Other Options Are Wrong

(A) This incorrect solution uses 100 mm/m instead of 1000 mm/m for particle size conversion. Other assumptions, definitions, and equations are the same as the correct solution.

$$v = \sqrt{\frac{(8)(0.05)\left(9.81 \frac{m}{s^2}\right)(0.22 \text{ mm}) \times \left(\frac{1 \text{ m}}{100 \text{ mm}}\right)(2.65 - 1)}{0.03}}$$

$$= 0.69 \text{ m/s}$$

$$w = \frac{Q}{dv}$$

$$= \frac{\left(3.5 \times 10^6 \frac{\text{gal}}{\text{day}}\right)\left(0.134 \frac{\text{ft}^3}{\text{gal}}\right)}{(4 \text{ ft})\left(0.69 \frac{m}{s}\right)\left(3.28 \frac{\text{ft}}{m}\right)\left(86{,}400 \frac{\text{sec}}{\text{day}}\right)}$$

$$= 0.60 \text{ ft}$$

(C) This incorrect solution fails to take the square root in the horizontal velocity calculation. Other assumptions, definitions, and equations are the same as the correct solution.

$$v = \frac{(8)(0.05)\left(9.81 \frac{m}{s^2}\right)(0.22 \text{ mm}) \times \left(\frac{1 \text{ m}}{1000 \text{ mm}}\right)(2.65 - 1)}{0.03}$$

$$= 0.047 \text{ m/s}$$

$$w = \frac{Q}{dv}$$

$$= \frac{\left(3.5 \times 10^6 \frac{\text{gal}}{\text{day}}\right)\left(0.134 \frac{\text{ft}^3}{\text{gal}}\right)}{(4 \text{ ft})\left(0.047 \frac{m}{s}\right)\left(3.28 \frac{\text{ft}}{m}\right)\left(86{,}400 \frac{\text{sec}}{\text{day}}\right)}$$

$$= 8.8 \text{ ft}$$

(D) This incorrect solution fails to convert gal to ft³ and assumes a hydraulic detention time. Other assumptions, definitions, and equations are the same as the correct solution.

$$v = \sqrt{\frac{(8)(0.05)\left(9.81 \frac{m}{s^2}\right)(0.22 \text{ mm}) \times \left(\frac{1 \text{ m}}{1000 \text{ mm}}\right)(2.65 - 1)}{0.03}}$$

$$= 0.22 \text{ m/s}$$

$$w = \frac{Q}{dv} = \frac{\left(3.5 \times 10^6 \frac{\text{gal}}{\text{day}}\right)\left(\frac{1 \text{ day}}{86{,}400 \text{ sec}}\right)}{(4 \text{ ft})\left(0.22 \frac{m}{s}\right)\left(3.28 \frac{\text{ft}}{m}\right)}$$

$$= 14 \text{ ft}$$

Units do not cancel to give feet.

SOLUTION 27

MLSS	mixed liquor suspended solids concentration	mg/L
SV	sludge volume	mL/L
SVI	sludge volume index	mL/g

$$\text{SVI} = \frac{\text{SV}}{\text{MLSS}} = \frac{\left(356 \ \frac{\text{mL}}{\text{L}}\right)\left(1000 \ \frac{\text{mg}}{\text{g}}\right)}{2400 \ \frac{\text{mg}}{\text{L}}}$$

$$= 148 \ \text{mL/g} \quad (150 \ \text{mL/g})$$

The answer is (A).

Why Other Options Are Wrong

(B) This incorrect solution uses the liquid volume as the sludge volume. Other assumptions, definitions, and equations are unchanged from the correct solution.

$$\text{SV} = 1000 \ \frac{\text{mL}}{\text{L}} - 356 \ \frac{\text{mL}}{\text{L}}$$
$$= 644 \ \text{mL/L}$$

$$\text{SVI} = \frac{\text{SV}}{\text{MLSS}} = \frac{\left(644 \ \frac{\text{mL}}{\text{L}}\right)\left(1000 \ \frac{\text{mg}}{\text{g}}\right)}{2400 \ \frac{\text{mg}}{\text{L}}}$$

$$= 268 \ \text{mL/g} \quad (270 \ \text{mL/g})$$

(C) This incorrect solution inverts SVI equation and uses liquid volume as the sludge volume. Other assumptions, definitions, and equations are the same as the correct solution.

$$\text{SV} = 1000 \ \frac{\text{mL}}{\text{L}} - 356 \ \frac{\text{mL}}{\text{L}}$$
$$= 644 \ \text{mL/L}$$

$$\text{SVI} = \frac{\text{MLSS}}{\text{SV}} = \frac{\left(2400 \ \frac{\text{mg}}{\text{L}}\right)\left(1000 \ \frac{\text{mg}}{\text{L}}\right)}{644 \ \frac{\text{mL}}{\text{L}}}$$

$$= 3727 \ \text{mL/g} \quad (3700 \ \text{mL/g})$$

(D) This incorrect solution inverts the SVI equation. Other assumptions and definitions are unchanged from the correct solution.

$$\text{SVI} = \frac{\text{MLSS}}{\text{SV}} = \frac{\left(2400 \ \frac{\text{mg}}{\text{L}}\right)\left(1000 \ \frac{\text{mg}}{\text{L}}\right)}{356 \ \frac{\text{mL}}{\text{L}}}$$

$$= 6742 \ \text{mL/g} \quad (6700 \ \text{mL/g})$$

Units are for concentration. Units of mL/g are assumed for SVI.

SOLUTION 28

C_L	chemical concentration corresponding to any future time	μg/L
C_o	chemical concentration t = 0 min	μg/L
$K_L a$	mass-transfer coefficient	min^{-1}
t	time	min

Under conditions of gas transfer from the liquid to the vapor phase, the following equation applies.

$$-K_L a = \frac{\ln \frac{C_o}{C_L}}{t}$$

elapsed time (min)	benzene concentration in sample (μg/L)	$\ln \frac{C_o}{C_L}$	mass-transfer coefficient (min^{-1})
0	978	–	–
2	569	0.542	−0.271
4	303	1.17	−0.293
6	153	1.86	−0.310
8	76	2.55	−0.319
10	36	3.30	−0.330
			−1.523

n = number of $K_L a$ calculations

$$-K_L a = \frac{\sum -K_L a_i}{n} = \frac{1.523 \ \text{min}^{-1}}{5}$$
$$= 0.30 \ \text{min}^{-1}$$

T	temperature	°C
θ	temperature correction coefficient	1.024 (typical)

$$-K_L a \text{ at } 8°C = (-K_L a \text{ at } 20°C)\theta^{T_2 - T_1}$$
$$= (0.30 \ \text{min}^{-1})(1.024^{8-20})$$
$$= 0.23 \ \text{min}^{-1}$$

The answer is (B).

Why Other Options Are Wrong

(A) This incorrect choice includes a math error by dividing by six (the number of concentration measurements) instead of by five (the number of calculations for $K_L a_i$). Other assumptions, definitions, and equations are unchanged from the correct solution.

$$-K_L a = \frac{\sum -K_L a_i}{n} = \frac{1.523 \ \text{min}^{-1}}{6}$$
$$= 0.25 \ \text{min}^{-1} \text{ at } 20°C$$

$$-K_L a \text{ at } 8°C = 0.25 \ \text{min}^{-1}(1.024^{8-20})$$
$$= 0.19 \ \text{min}^{-1}$$

(C) This incorrect choice does not correct for temperature. Other assumptions, definitions, and equations are the same as for the correct solution.

$$-K_L a = \frac{\sum -K_L a_i}{n}$$
$$= \frac{1.523 \text{ min}^{-1}}{5}$$
$$= 0.30 \text{ min}^{-1}$$

(D) This incorrect choice improperly corrects for temperature. Other assumptions, definitions, and equations are unchanged from the correct solution.

$$-K_L a = \frac{\sum -K_L a_i}{n}$$
$$= \frac{1.523 \text{ min}^{-1}}{5}$$
$$= 0.30 \text{ min}^{-1} \text{ at } 20°C$$
$$-K_L a \text{ at } 8°C = (-K_L a \text{ at } 20°C)\theta^{T_1-T_2}$$
$$= (0.30 \text{ min}^{-1})(1.024^{20-8})$$
$$= 0.40 \text{ min}^{-1}$$

SOLUTION 29

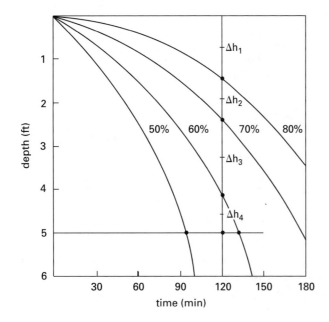

D	total depth	ft
E	total efficiency	%
R_i	incremental efficiency	%
Δh_i	incremental change in depth	ft

$$E = \sum 0.5 \left(\frac{\Delta h_i}{d}\right)(R_i + R_{i+1})$$
$$= (0.5)\left(\frac{1.42 \text{ ft}}{5 \text{ ft}}\right)(100\% + 80\%)$$
$$+ (0.5)\left(\frac{0.95 \text{ ft}}{5 \text{ ft}}\right)(80\% + 70\%)$$
$$+ (0.5)\left(\frac{1.68 \text{ ft}}{5 \text{ ft}}\right)(70\% + 60\%)$$
$$+ (0.5)\left(\frac{0.82 \text{ ft}}{5 \text{ ft}}\right)(60\% + 57\%)$$
$$= 25.56\% + 14.25\% + 21.84\% + 9.59\%$$
$$= 71\%$$

The answer is (D).

Why Other Options Are Wrong

(A) This incorrect solution uses the efficiency at the height-time coordinate of 5 ft and 120 min subtracted from 100%.

$$\text{efficiency} = 100\% - 57\% = 43\%$$

(B) This incorrect solution uses the efficiency as the height-time coordinate for 5 ft and 120 min.

$$\text{efficiency} = 57\%$$

(C) This incorrect solution does not average the incremental efficiencies. The illustration and definitions are unchanged from the correct solution.

$$E = \sum \frac{\Delta h_i R_i}{d}$$
$$= \frac{1.42 \text{ ft}}{5 \text{ ft}} \times 80\% + \frac{0.95 \text{ ft}}{5 \text{ ft}} \times 70\%$$
$$+ \frac{1.68 \text{ ft}}{5 \text{ ft}} \times 60\% + \frac{0.82 \text{ ft}}{5 \text{ ft}} \times 57\%$$
$$= 22.72\% + 13.30\% + 20.16\% + 9.35\%$$
$$= 66\%$$

SOLUTION 30

The winter loading rate will control because it results in the largest surface area.

C	BOD concentration	mg/L
$\dot{V}$	volumetric flow rate	gal/day
$\dot{m}$	mass flow rate	lbm/day

$$\dot{m} = \dot{V}C$$
$$= \left(2.6 \times 10^5 \frac{\text{gal}}{\text{day}}\right)\left(14\,000 \frac{\text{mg BOD}}{\text{L}}\right)$$
$$\times \left(3.785 \frac{\text{L}}{\text{gal}}\right)\left(2.204 \frac{\text{lbm}}{10^6 \text{ mg}}\right)$$
$$= 30{,}365 \text{ lbm BOD/day}$$

DEPTH SOLUTIONS

A surface area   ac
D liquid depth   ft
OLR   organic loading rate   lbm/10^3 ft^3-day

$$A = \frac{\dot{m}}{(\text{OLR})D}$$

$$= \frac{\left(30{,}365 \, \frac{\text{lbm BOD}}{\text{day}}\right)\left(\frac{1 \text{ ac}}{43{,}560 \text{ ft}^2}\right)}{\left(12 \, \frac{\text{lbm BOD}}{10^3 \text{ ft}^3\text{-day}}\right)(10 \text{ ft})}$$

$$= 5.8 \text{ ac}$$

The answer is (D).

Why Other Options Are Wrong

(A) This incorrect solution calculates the surface area based on summer loading rate. Other definitions and equations are unchanged from the correct solution.

Assume that the summer loading rate will control because it is larger than the winter loading rate.

$$\dot{m} = \left(2.6 \times 10^5 \, \frac{\text{gal}}{\text{day}}\right)\left(14\,000 \, \frac{\text{mg BOD}}{\text{L}}\right)$$
$$\times \left(3.785 \, \frac{\text{L}}{\text{gal}}\right)\left(2.204 \, \frac{\text{lbm}}{10^6 \text{ mg}}\right)$$
$$= 30{,}365 \text{ lbm BOD/day}$$

$$A = \frac{\left(30{,}365 \, \frac{\text{lbm BOD}}{\text{day}}\right)\left(\frac{1 \text{ ac}}{43{,}560 \text{ ft}^2}\right)}{\left(18 \, \frac{\text{lbm BOD}}{10^3 \text{ ft}^3\text{-day}}\right)(10 \text{ ft})}$$

$$= 3.9 \text{ ac}$$

(B) This incorrect solution calculates the surface area based on the average of the winter and summer loading rates.

$$\dot{m} = \left(2.6 \times 10^5 \, \frac{\text{gal}}{\text{day}}\right)\left(14\,000 \, \frac{\text{mg BOD}}{\text{L}}\right)$$
$$\times \left(3.785 \, \frac{\text{L}}{\text{gal}}\right)\left(2.204 \, \frac{\text{lbm}}{10^6 \text{ mg}}\right)$$
$$= 30{,}365 \text{ lbm BOD/day}$$

OLR_{mean} average organic loading rate   lbm BOD/10^3 ft^3
$\text{OLR}_{\text{winter}}$ organic loading rate during winter months   lbm BOD/10^3 ft^3
$\text{OLR}_{\text{summer}}$ average organic loading rate during summer months   lbm BOD/10^3 ft^3

$$\text{OLR}_{\text{mean}} = \frac{\text{OLR}_{\text{winter}} + \text{OLR}_{\text{summer}}}{2}$$
$$= \frac{12 \, \frac{\text{lbm BOD}}{10^3 \text{ ft}^3} + 18 \, \frac{\text{lbm BOD}}{10^3 \text{ ft}^3}}{2}$$
$$= 15 \text{ lbm BOD}/10^3 \text{ ft}^3$$

$$A = \frac{Q\dot{m}}{(\text{OLR}_{\text{mean}})D}$$

$$= \frac{\left(30{,}365 \, \frac{\text{lbm BOD}}{\text{day}}\right)\left(\frac{1 \text{ ac}}{43{,}560 \text{ ft}^2}\right)}{\left(15 \, \frac{\text{lbm BOD}}{10^3 \text{ ft}^3\text{-day}}\right)(10 \text{ ft})}$$

$$= 4.6 \text{ ac}$$

(C) This incorrect solution includes an error in the conversion factor from liters to gallons. Other definitions and equations are unchanged from the correct solution.

Assume that the winter loading rate will control because it provides the largest surface area.

$$\dot{m} = \left(2.6 \times 10^5 \, \frac{\text{gal}}{\text{day}}\right)\left(14\,000 \, \frac{\text{mg BOD}}{\text{L}}\right)$$
$$\times \left(3.28 \, \frac{\text{L}}{\text{gal}}\right)\left(2.204 \, \frac{\text{lbm}}{10^6 \text{ mg}}\right)$$
$$= 26{,}314 \text{ lbm BOD/day}$$

$$A = \frac{\left(26{,}314 \, \frac{\text{lbm BOD}}{\text{day}}\right)\left(\frac{1 \text{ ac}}{43{,}560 \text{ ft}^2}\right)}{\left(12 \, \frac{\text{lbm BOD}}{10^3 \text{ ft}^3\text{-day}}\right)(10 \text{ ft})}$$

$$= 5.0 \text{ ac}$$

SOLUTION 31

E permeate recovery fraction   –
Q_d desired flow rate   m^3/d
Q_r feed flow rate   m^3/d

$$Q_d = \frac{Q_E}{E} = \frac{16\,000 \, \frac{\text{m}^3}{\text{d}}}{0.80}$$
$$= 20\,000 \text{ m}^3/\text{d}$$

G_m membrane flux rate   m^3/m^2·d
V_m membrane volume   m^3
ρ_m membrane packing density   m^2/m^3

$$V_m = \frac{Q_d}{G_m \rho_m} = \frac{20\,000 \; \frac{m^3}{d}}{\left(0.83 \; \frac{m^3}{m^2 \cdot d}\right)\left(800 \; \frac{m^2}{m^3}\right)}$$

$$= 30 \; m^3$$

The number of pressure vessels is

$$(30 \; m^3)\left(\frac{1 \; module}{0.03 \; m^3}\right)\left(\frac{1 \; pressure \; vessel}{10 \; modules}\right)$$

$$= 100 \; pressure \; vessels$$

The answer is (D).

Why Other Options Are Wrong

(A) This incorrect choice multiplies instead of divides the desired freshwater flow rate by the permeate recovery. Other definitions and equations are unchanged from the correct solution.

$$Q_d = Q_r E = \left(16\,000 \; \frac{m^3}{d}\right)(0.80)$$

$$= 12\,800 \; m^3/d$$

$$V_m = \frac{12\,800 \; \frac{m^3}{d}}{\left(0.83 \; \frac{m^3}{m^2 \cdot d}\right)\left(800 \; \frac{m^2}{m^3}\right)} = 19 \; m^3$$

The number of pressure vessels is

$$(19 \; m^3)\left(\frac{1 \; module}{0.03 \; m^3}\right)\left(\frac{1 \; pressure \; vessel}{10 \; modules}\right) = 63$$

(B) This incorrect choice uses the desired freshwater flow rate instead of calculating the feed rate corrected for permeate recovery. Other definitions and equations are unchanged from the correct solution.

$$V_m = \frac{16\,000 \; \frac{m^3}{d}}{\left(0.83 \; \frac{m^3}{m^2 \cdot d}\right)\left(800 \; \frac{m^2}{m^3}\right)} = 24 \; m^3$$

The number of pressure vessels is

$$(24 \; m^3)\left(\frac{1 \; module}{0.03 \; m^3}\right)\left(\frac{1 \; pressure \; vessel}{10 \; modules}\right) = 80$$

(C) This incorrect choice uses the salt rejection percentage instead of the permeate recovery percentage to find the feed rate. Other definitions and equations are unchanged from the correct solution.

$$Q_d = \frac{16\,000 \; \frac{m^3}{d}}{0.92}$$

$$= 17\,391 \; m^3/d$$

$$V_m = \frac{17\,391 \; \frac{m^3}{d}}{\left(0.83 \; \frac{m^3}{m^2 \cdot d}\right)\left(800 \; \frac{m^2}{m^3}\right)}$$

$$= 26 \; m^3$$

The number of pressure vessels is

$$(26 \; m^3)\left(\frac{1 \; module}{0.03 \; m^3}\right)\left(\frac{1 \; pressure \; vessel}{10 \; modules}\right) = 87$$

SOLUTION 32

f	solids fraction	%
$\dot{m}$	dry solids mass flow rate	kg/d
V	wet sludge volumetric flow rate	m^3/d
ρ	sludge density	kg/m^3

Assume that for a sludge at 9% solids, the sludge density is equal to that of water ($1000 \; kg/m^3$).

$$\dot{m} = V\rho f$$

$$= \left(18{,}000 \; \frac{gal}{day}\right)\left(1000 \; \frac{kg}{m^3}\right)(0.09)$$

$$\times \left(0.003\,785 \; \frac{m^3}{gal}\right)$$

$$= 6132 \; kg/d$$

The lime requirement is

$$\frac{\left(6132 \; \frac{kg}{d}\right)\left(30 \; \frac{d}{month}\right)\left(315 \; g \frac{Ca(OH)_2}{kg}\right)}{(1 - 0.22)\left(1000 \; \frac{g}{kg}\right)}$$

$$= 74\,292 \; kg/month \quad (74\,000 \; kg/month)$$

The answer is (C).

Why Other Options Are Wrong

(A) This incorrect solution corrects for the inert materials in the lime by multiplying instead of dividing by the impurity fraction. Other assumptions, definitions, and equations are the same as the correct solution.

$$\dot{m} = \left(18{,}000 \ \frac{\text{gal}}{\text{day}}\right)\left(1000 \ \frac{\text{kg}}{\text{m}^3}\right)(0.09)$$
$$\times \left(0.003\,785 \ \frac{\text{m}^3}{\text{gal}}\right)$$
$$= 6132 \ \text{kg/d}$$

The lime requirement is

$$\frac{\left(6132 \ \frac{\text{kg}}{\text{d}}\right)\left(30 \ \frac{\text{d}}{\text{month}}\right)}{1000 \ \frac{\text{g}}{\text{kg}}}$$
$$\times \left(315 \ \text{g} \ \frac{\text{Ca(OH)}_2}{\text{kg}}\right)(1-0.22)$$
$$= 45\,199 \ \text{kg/month} \quad (45\,000 \ \text{kg/month})$$

(B) This incorrect solution does not correct for inert materials in the lime. Other assumptions, definitions, and equations are the same as the correct solution.

$$\dot{m} = \left(18{,}000 \ \frac{\text{gal}}{\text{day}}\right)\left(1000 \ \frac{\text{kg}}{\text{m}^3}\right)(0.09)$$
$$\times \left(0.003\,785 \ \frac{\text{m}^3}{\text{gal}}\right)$$
$$= 6132 \ \text{kg/d}$$

The lime requirement is

$$\frac{\left(6132 \ \frac{\text{kg}}{\text{d}}\right)\left(30 \ \frac{\text{d}}{\text{month}}\right)\left(315 \ \text{g} \ \frac{\text{Ca(OH)}_2}{\text{kg}}\right)}{1000 \ \frac{\text{g}}{\text{kg}}}$$
$$= 57\,947 \ \text{kg/month} \quad (58\,000 \ \text{kg/month})$$

(D) This incorrect solution uses the solids fraction for dewatered solids instead of for solids as wasted. Other assumptions, definitions, and equations are unchanged from the correct solution.

$$\dot{m} = \left(18{,}000 \ \frac{\text{gal}}{\text{day}}\right)\left(1000 \ \frac{\text{kg}}{\text{m}^3}\right)(0.23)$$
$$\times \left(0.003\,785 \ \frac{\text{m}^3}{\text{gal}}\right)$$
$$= 15\,670 \ \text{kg/d}$$

The lime requirement is

$$\frac{\left(15\,670 \ \frac{\text{kg}}{\text{d}}\right)\left(30 \ \frac{\text{d}}{\text{month}}\right)\left(315 \ \text{g} \ \frac{\text{Ca(OH)}_2}{\text{kg}}\right)}{(1-0.22)\left(1000 \ \frac{\text{g}}{\text{kg}}\right)}$$
$$= 189\,848 \ \text{kg/month} \quad (190\,000 \ \text{kg/month})$$

SOLUTION 33

Q_w	wasted solids flow rate	m³/d
X_u	wasted solids concentration	mg/L

The wasted sludge mass is

$$Q_w X_u = 34 \ \text{kg/d}$$

V	reactor volume	m³
X	biomass concentration in the reactor	mg/L
θ_c	mean cell residence time	d

$$\theta_c = \frac{VX}{Q_w X_u}$$

The bioreactor sludge mass is

$$VX = \theta_c(Q_w X_u) = (10 \ \text{d})\left(34 \ \frac{\text{kg}}{\text{d}}\right)$$
$$= 340 \ \text{kg}$$

F/M	food-to-microorganism ratio	d⁻¹
Q	influent wastewater flow rate	m³/d
S_o	influent wastewater BOD	mg/L

$$\frac{F}{M} = \frac{QS_o}{VX}$$

$$= \frac{\left(18\,925 \ \frac{\text{m}^3}{\text{d}}\right)\left(247 \ \frac{\text{mg}}{\text{L}}\right)}{340 \ \text{kg}}$$
$$\times \left(1000 \ \frac{\text{L}}{\text{m}^3}\right)\left(10^{-6} \ \frac{\text{kg}}{\text{mg}}\right)$$
$$= 13.7 \ \text{d}^{-1} \quad (14 \ \text{d}^{-1})$$

The answer is (B).

Why Other Options Are Wrong

(A) This incorrect solution fails to apply all necessary conversion factors required for consistent units. Other definitions and equations are the same as used in the correct solution.

$$Q_w X_u = 34 \ \text{kg/d}$$

$$VX = (10 \text{ d})\left(34 \ \frac{\text{kg}}{\text{d}}\right)$$
$$= 340 \text{ kg}$$

$$\frac{F}{M} = \frac{\left(18925 \ \frac{\text{m}^3}{\text{d}}\right)\left(247 \ \frac{\text{mg}}{\text{L}}\right) \times \left(10^{-6} \ \frac{\text{kg}}{\text{mg}}\right)}{340 \text{ kg}}$$
$$= 0.014 \text{ d}^{-1}$$

Units do not work.

(C) This incorrect solution uses the wasted sludge mass as the bioreactor sludge mass. Other definitions and equations are the same as used in the correct solution.

$$Q_w X_u = VX$$
$$= 34 \text{ kg/d}$$

$$\frac{F}{M} = \frac{\left(18925 \ \frac{\text{m}^3}{\text{d}}\right)\left(247 \ \frac{\text{mg}}{\text{L}}\right) \times \left(1000 \ \frac{\text{L}}{\text{m}^3}\right)\left(10^{-6} \ \frac{\text{kg}}{\text{mg}}\right)}{34 \ \frac{\text{kg}}{\text{d}}}$$
$$= 137 \text{ d}^{-1} \quad (140 \text{ d}^{-1})$$

Units do not work.

(D) This incorrect solution improperly defines the mean cell residence time and makes an error in applying a conversion factor. Other definitions and equations are the same as used in the correct solution.

$$Q_w X_u = \text{wasted sludge mass}$$
$$= 34 \text{ kg/d}$$
$$\theta_c = Q_w X_u / VX$$
$$VX = \frac{Q_w X_u}{\theta_c} = \frac{34 \ \frac{\text{kg}}{\text{d}}}{10 \text{ d}}$$
$$= 3.4 \text{ kg}$$

$$\frac{F}{M} = \frac{\left(18925 \ \frac{\text{m}^3}{\text{d}}\right)\left(247 \ \frac{\text{mg}}{\text{L}}\right) \times \left(1000 \ \frac{\text{L}}{\text{m}^3}\right)\left(10^{-6} \ \frac{\text{kg}}{\text{mg}}\right)}{3.4 \text{ kg}}$$
$$= 1375 \text{ d}^{-1} \quad (1400 \text{ d}^{-1})$$

Units do not work.

SOLUTION 34

k_d	endogenous decay rate constant	d^{-1}
Y	yield coefficient	mg/mg
Y_{obs}	observed yield coefficient	mg/mg
θ_c	mean cell residence time	d

$$Y_{\text{obs}} = \frac{Y}{1 + k_d \theta_c} = \frac{0.53 \ \frac{\text{g}}{\text{g}}}{1 + \left(\frac{0.05}{\text{d}}\right)(8 \text{ d})}$$
$$= 0.38 \text{ g/g}$$

Q	influent flow rate	m^3/d
S	effluent BOD	mg/L
S_o	influent BOD	mg/L
X_p	mass of biomass produced	kg/d

$$X_p = Y_{\text{obs}}(S_o - S)Q$$
$$= \left(0.38 \ \frac{\text{g}}{\text{g}}\right)\left(281 \ \frac{\text{mg}}{\text{L}} - 20 \ \frac{\text{mg}}{\text{L}}\right)\left(27000 \ \frac{\text{m}^3}{\text{d}}\right)$$
$$\times \left(1000 \ \frac{\text{L}}{\text{m}^3}\right)\left(10^{-6} \ \frac{\text{kg}}{\text{mg}}\right)$$
$$= 2678 \text{ kg/d} \quad (2700 \text{ kg/d})$$

The answer is (A).

Why Other Options Are Wrong

(B) This incorrect choice does not correct the yield coefficient for cell death. Other equations are unchanged from the correct solution.

$$X_p = Y(S_o - S)Q$$
$$= \left(0.53 \ \frac{\text{g}}{\text{g}}\right)\left(281 \ \frac{\text{mg}}{\text{L}} - 20 \ \frac{\text{mg}}{\text{L}}\right)\left(27000 \ \frac{\text{m}^3}{\text{d}}\right)$$
$$\times \left(1000 \ \frac{\text{L}}{\text{m}^3}\right)\left(10^{-6} \ \frac{\text{kg}}{\text{mg}}\right)$$
$$= 3735 \text{ kg/d} \quad (3700 \text{ kg/d})$$

(C) This incorrect choice does not include the yield coefficient in the calculation. Other definitions are unchanged from the correct solution.

$$X_p = (S_o - S)Q$$
$$= \left(281 \ \frac{\text{mg}}{\text{L}} - 20 \ \frac{\text{mg}}{\text{L}}\right)\left(27000 \ \frac{\text{m}^3}{\text{d}}\right)$$
$$\times \left(1000 \ \frac{\text{L}}{\text{m}^3}\right)\left(10^{-6} \ \frac{\text{kg}}{\text{mg}}\right)$$
$$= 7047 \text{ kg/d} \quad (7000 \text{ kg/d})$$

(D) This incorrect choice miscalculates the observed yield coefficient. Other definitions and equations are unchanged from the correct solution.

$$Y_{obs} = 1 + \frac{Y}{k_d \theta_c} = 1 + \frac{0.53 \frac{g}{g}}{\left(0.05 \frac{1}{d}\right)(8\text{ d})}$$

$$= 2.33 \text{ g/g}$$

$$X_p = \left(2.33 \frac{g}{g}\right)\left(281 \frac{mg}{L} - 20 \frac{mg}{L}\right)\left(27\,000 \frac{m^3}{d}\right)$$
$$\times \left(1000 \frac{L}{m^3}\right)\left(10^{-6} \frac{kg}{mg}\right)$$
$$= 16\,420 \text{ kg/d} \quad (16\,000 \text{ kg/d})$$

SOLUTION 35

Check the surface area required based on hydraulic loading. Typical hydraulic loading rates vary between 2.0 and 4.0 gal/ft²-day. To be conservative, use the minimum value of 2.0 gal/ft²-day.

A_h	media surface area based on hydraulic loading	ft²
HLR	hydraulic loading rate	gal/ft²-day
Q	influent flow rate	gal/day

$$A_h = \frac{Q}{\text{HLR}} = \frac{250{,}000 \frac{\text{gal}}{\text{day}}}{2.0 \frac{\text{gal}}{\text{ft}^2-\text{day}}}$$
$$= 125{,}000 \text{ ft}^2$$

Check the surface area required based on organic loading. Typical organic loading rates vary between 2.0 and 3.5 lbm total BOD/10³ ft²-day. To be conservative, use the minimum value of 2.0 lbm/10³ ft²-day.

A_s	media surface area based on organic loading	ft²
OLR	organic loading rate	lbm/10³ ft²-day
S_o	influent total BOD	mg/L

$$A_s = \frac{QS_o}{\text{OLR}}$$

$$= \frac{\left(250{,}000 \frac{\text{gal}}{\text{day}}\right)\left(174 \frac{\text{mg}}{\text{L}}\right)}{2.0 \frac{\text{lbm}}{10^3 \text{ ft}^2-\text{day}}}$$

$$= 181{,}442 \text{ ft}^2 \quad (180{,}000 \text{ ft}^2)$$

$$A_s > A_h$$

Total media surface area is 180,000 ft².

The answer is (C).

Why Other Options Are Wrong

(A) This incorrect option does not check surface area based on organic loading rate. Other assumptions, definitions, and equations are the same as those used in the correct solution.

$$A_h = \frac{250{,}000 \frac{\text{gal}}{\text{day}}}{2.0 \frac{\text{gal}}{\text{ft}^2-\text{day}}}$$
$$= 125{,}000 \text{ ft}^2 \quad (130{,}000 \text{ ft}^2)$$

The media surface area is 130,000 ft².

(B) This incorrect option subtracts the effluent BOD from the influent BOD in the calculation for surface area based on organic loading. Other assumptions, definitions, and equations are the same as those used in the correct solution.

$$A_h = \frac{250{,}000 \frac{\text{gal}}{\text{day}}}{2.0 \frac{\text{gal}}{\text{ft}^2-\text{day}}} = 125{,}000 \text{ ft}^2$$

$$A_s = \frac{\left(250{,}000 \frac{\text{gal}}{\text{day}}\right)\left(174 \frac{\text{mg}}{\text{L}} - 30 \frac{\text{mg}}{\text{L}}\right)}{2.0 \frac{\text{lbm}}{10^3 \text{ ft}^2-\text{day}}}$$

$$= 150{,}159 \text{ ft}^2 \quad (150{,}000 \text{ ft}^2)$$

$$A_s > A_h$$

Total media surface area is 150,000 ft².

(D) This incorrect option includes the wrong conversion factor for consistent volume units. Other assumptions, definitions, and equations are the same as used in the correct solution.

$$A_h = \frac{250{,}000 \frac{\text{gal}}{\text{day}}}{2.0 \frac{\text{gal}}{\text{ft}^2-\text{day}}} = 125{,}000 \text{ ft}^2$$

$$A_s = \frac{\left(250{,}000 \frac{\text{gal}}{\text{day}}\right)\left(174 \frac{\text{mg}}{\text{L}}\right)}{2.0 \frac{\text{lbm}}{10^3 \text{ ft}^2-\text{day}}}$$

$$= 357{,}610 \text{ ft}^2 \quad (360{,}000 \text{ ft}^2)$$

Units do not work.

$$A_s > A_h$$

Total media surface area is 360,000 ft².

SOLUTION 36

An SF wetland system may be selected over an FWS wetland system for reasons of improved odor and vector control, reduced public exposure, improved suspended solids removal, lower susceptibility to climatic temperature extremes, and greater reaction rates. However, because the exposed water surface provides a better opportunity for oxygen transfer, the FWS system may be preferred where an aerobic environment is necessary to effect a desired treatment such as nitrification for ammonia removal.

The answer is (C).

Why Other Options Are Wrong

(A) This choice is incorrect because the SF wetland is operated without exposing the water surface. The submerged flow limits vector access and odor dispersion by wind currents.

(B) This choice is incorrect because the SF wetland provides an opportunity for filtration as the water passes through the media and, because the water is not exposed to sunlight, reduces the occurrence of algae in the effluent. Both of these contribute to improved suspended solids removal.

(D) This choice is incorrect because by not exposing the water surface to the air, the SF wetland experiences less impact from ambient air temperature fluctuations and other climatic variations.

SOLUTION 37

A/S	air-to-solids ratio	ft^3/lbm
C	solids concentration	mg/L
V_a	air volumetric flow rate	gal/day
Q_s	suspension flow rate	gal/day

$$\dot{V}_a = \frac{A}{S} C Q_s$$
$$= \left(1.3 \ \frac{ft^3}{lbm}\right) \left(1200 \ \frac{mg}{L}\right) \left(2.204 \ \frac{lbm}{10^6 \ mg}\right)$$
$$\times \left(3.785 \ \frac{L}{gal}\right) \left(5.0 \times 10^5 \ \frac{gal}{day}\right)$$
$$= 6507 \ ft^3/day$$

$\dot{m}_a$	air mass flow rate	lbm/min
ρ_a	air density lbm/ft^3	

At 80°F and 1 atm, ρ_a is 0.0735 lbm/ft^3.

$$\dot{m}_a = \dot{V}_a \rho_a$$
$$= \left(6507 \ \frac{ft^3}{day}\right) \left(0.0735 \ \frac{lbm}{ft^3}\right)$$
$$= 478 \ lbm/day \quad (480 \ lbm/day)$$

The answer is (A).

Why Other Options Are Wrong

(B) This incorrect solution divides by water density, instead of multiplying, in the air flow rate calculation. The units do not work in this calculation. Other definitions and equations are the same as used in the correct solution.

$$\dot{V}_a = \left(1.3 \ \frac{ft^3}{lbm}\right) \left(1200 \ \frac{mg}{L}\right) \left(2.204 \ \frac{lbm}{10^6 \ mg}\right)$$
$$\times \left(3.785 \ \frac{L}{gal}\right) \left(5.0 \times 10^5 \ \frac{gal}{day}\right)$$
$$= 6507 \ ft^3/day$$

$$\dot{m}_a = \frac{Q_a}{\rho_a} = \frac{6507 \ \frac{ft^3}{day}}{0.0735 \ \frac{lbm}{ft^3}}$$
$$= 8.9 \times 10^4 \ lbm/day$$

Note that units do not work.

(C) This incorrect solution uses water density instead of air density in the air mass flow equation. Other definitions and equations are the same as used in the correct solution.

$$\dot{V}_a = \left(1.3 \ \frac{ft^3}{lbm}\right) \left(1200 \ \frac{mg}{L}\right) \left(2.204 \ \frac{lbm}{10^6 \ mg}\right)$$
$$\times \left(3.785 \ \frac{L}{gal}\right) \left(5.0 \times 10^5 \ \frac{gal}{day}\right)$$
$$= 6507 \ ft^3/day$$

At 80°F and 1 atm, ρ_a is 62 lbm/ft^3.

$$\dot{m}_a = \left(6507 \ \frac{ft^3}{day}\right) \left(62 \ \frac{lbm}{ft^3}\right)$$
$$= 4.0 \times 10^5 \ lbm/day$$

(D) This incorrect solution does not include the suspension concentration in the air flow rate calculation and uses water density to obtain consistent units. Other definitions and equations are the same as used in the correct solution.

ρ_w = water density = 62 lbm/ft^3

$$\dot{V}_a = \left(\frac{A}{S}\right)(Q_s)(\rho_w)$$

$$= \left(1.3 \frac{\text{ft}^3}{\text{lbm}}\right)\left(5.0 \times 10^5 \frac{\text{gal}}{\text{day}}\right)$$

$$\times \left(62 \frac{\text{lbm}}{\text{ft}^3}\right)\left(0.134 \frac{\text{ft}^3}{\text{gal}}\right)$$

$$= 5.4 \times 10^6 \text{ ft}^3/\text{day}$$

$$\dot{m}_a = \left(5.4 \times 10^6 \frac{\text{ft}^3}{\text{day}}\right)\left(62 \frac{\text{lbm}}{\text{ft}^3}\right)$$

$$= 3.3 \times 10^8 \text{ lbm/day}$$

SOLUTION 38

For a complete mix reactor without recycle, the hydraulic residence time and the mean cell residence time are equal.

k_d	endogenous decay rate coefficient	d^{-1}
S	effluent biochemical oxygen demand (BOD)	mg/L
S_o	influent BOD	mg/L
X	mixed liquor suspended solids (MLSS)	mg/L
Y	yield coefficient	g/g
θ	hydraulic residence time	d
θ_c	mean cell residence time	d

Values for the yield coefficient and the endogenous decay-rate coefficient are found from the following linear equation. The yield coefficient is the slope and the endogenous decay rate coefficient is the intercept.

$$\frac{1}{\theta_c} = \frac{Y(S_o - S)}{\theta X} - k_d$$

$1/\theta_c$ (d^{-1})	$(S_o - S)/\theta X$ (d^{-1})	equation
0.29	0.49	$0.29 = Y\,0.49 - k_d$
0.52	0.75	$0.52 = Y\,0.75 - k_d$
0.69	0.95	$0.69 = Y\,0.95 - k_d$
0.91	1.2	$0.91 = Y\,1.2 - k_d$

Solving any two of the equations simultaneously will give values for the yield coefficient and the endogenous decay-rate coefficient.

$$\frac{0.29}{\text{d}} = Y\frac{0.49}{\text{d}} - k_d$$

$$Y = \frac{\frac{0.29}{\text{d}} + k_d}{\frac{0.49}{\text{d}}}$$

$$\frac{0.69}{\text{d}} = \frac{\left(\frac{0.29}{\text{d}} + k_d\right)\left(\frac{0.95}{\text{d}}\right)}{\frac{0.49}{\text{d}}} - k_d$$

$$= \frac{\left(\frac{0.95}{\text{d}}\right)\left(\frac{0.29}{\text{d}}\right)}{\frac{0.49}{\text{d}}} + \frac{\frac{0.95}{\text{d}}k_d}{\frac{0.49}{\text{d}}} - k_d$$

$$= \frac{0.56}{\text{d}} + \frac{1.9}{\text{d}}k_d - k_d$$

$$k_d = \frac{\frac{0.69}{\text{d}} - \frac{0.56}{\text{d}}}{\frac{0.9}{\text{d}}}$$

$$= 0.14 \text{ d}^{-1}$$

The answer is (C).

Why Other Options Are Wrong

(A) This incorrect solution makes a mathematical error when solving the two equations. Other assumptions, definitions, and equations are unchanged from the correct solution.

$1/\theta_c$ (d^{-1})	$(S_o - S)/\theta X$ (d^{-1})	equation
0.29	0.49	$0.29 = Y\,0.49 - k_d$
0.52	0.75	$0.52 = Y\,0.75 - k_d$
0.69	0.95	$0.69 = Y\,0.95 - k_d$
0.91	1.2	$0.91 = Y\,1.2 - k_d$

$$\frac{0.29}{\text{d}} = Y\frac{0.49}{\text{d}} - k_d$$

$$Y = \frac{\frac{0.29}{\text{d}}}{\frac{0.49}{\text{d}}} + k_d$$

$$\frac{0.69}{\text{d}} = \frac{\left(\frac{0.29}{\text{d}} + k_d\right)\left(\frac{0.95}{\text{d}}\right)}{\frac{0.49}{\text{d}}} - k_d$$

$$\left(\frac{0.69}{\text{d}}\right)\left(\frac{0.49}{\text{d}}\right) = \left(\frac{0.29}{\text{d}}\right)\left(\frac{0.95}{\text{d}}\right) + \frac{0.95}{\text{d}}$$
$$\times k_d - k_d$$

$$\frac{0.34}{\text{d}} = \frac{0.28}{\text{d}} - \frac{0.05}{\text{d}}k_d$$

$$k_d = \frac{\frac{0.34}{\text{d}} - \frac{0.28}{\text{d}}}{-\frac{0.05}{\text{d}}}$$

$$= -1.2 \text{ d}^{-1}$$

(B) This incorrect solution assumes a mean cell residence time of 10 d. Other definitions and equations are unchanged from the correct solution.

Since the mean cell residence time is not given, assume 10 d as a typical value.

$1/\theta_c$ (d^{-1})	$(S_o - S)/\theta X$ (d^{-1})	equation
0.1	0.49	$0.1 = Y 0.49 - k_d$
0.1	0.75	$0.1 = Y 0.75 - k_d$
0.1	0.95	$0.1 = Y 0.95 - k_d$
0.1	1.2	$0.1 = Y 1.2 - k_d$

Assume the equations all define points on the same straight line and solve any two to find the endogenous decay rate coefficient.

$$\frac{0.1}{d} = Y\frac{0.49}{d} - k_d$$

$$Y = \frac{\frac{0.1}{d} + k_d}{\frac{0.49}{d}}$$

$$\frac{0.1}{d} = \frac{\left(\frac{0.1}{d} + k_d\right)\left(\frac{0.95}{d}\right)}{\frac{0.49}{d}} - k_d$$

$$= \frac{\left(\frac{0.95}{d}\right)\left(\frac{0.1}{d}\right)}{\frac{0.49}{d}} + \frac{\frac{0.95}{d}k_d}{\frac{0.49}{d}} - k_d$$

$$= \frac{0.19}{d} + \frac{1.9}{d}k_d - k_d$$

$$k_d = \frac{\frac{0.1}{d} - \frac{0.19}{d}}{\frac{0.9}{d}}$$

$$= -0.10 \text{ d}^{-1}$$

(D) This incorrect solution solves for the yield coefficient. Other assumptions, definitions, and equations are unchanged from the correct solution.

$1/\theta_c$ (d^{-1})	$(S_o - S)/\theta X$ (d^{-1})	equation
0.29	0.49	$0.29 = Y 0.49 - k_d$
0.52	0.75	$0.52 = Y 0.75 - k_d$
0.69	0.95	$0.69 = Y 0.95 - k_d$
0.91	1.2	$0.91 = Y 1.2 - k_d$

$$\frac{0.29}{d} = Y\frac{0.49}{d} - k_d$$

$$k_d = Y\frac{0.49}{d} - \frac{0.29}{d}$$

$$\frac{0.69}{d} = Y\frac{0.95}{d} - Y\frac{0.49}{d} + \frac{0.29}{d}$$

$$Y = \frac{\frac{0.69}{d} - \frac{0.29}{d}}{\frac{0.95}{d} - \frac{0.49}{d}}$$

$$= 0.87$$

To get proper units, let k_d be 0.87 d^{-1}.

SOLUTION 39

k	growth rate	d^{-1}
Y	yield coefficient	g/g
μ_{mD}	maximum growth rate for denitrification	d^{-1}
μ'_{mD}	corrected maximum growth rate for denitrification	d^{-1}

$$\mu_{mD} = kY = \left(\frac{0.38}{d}\right)\left(0.81 \frac{g}{g}\right) = 0.31 \text{ d}^{-1}$$

C_N nitrate and/or nitrite concentration mg/L

$$C_N = 29 \frac{\text{mg}}{\text{L}} + 8 \frac{\text{mg}}{\text{L}} = 37 \text{ mg/L}$$

K_M	half velocity constant for methanol	mg/L
K_N	half velocity constants for nitrogen	mg/L
C_M	methanol concentration	mg/L
T	temperature	°C

$$\mu'_{mD} = \mu_{mD} 0.0025 T^2 \left(\frac{C_M}{K_M + C_M}\right)\left(\frac{C_N}{K_N + C_N}\right)$$

$$= \left(\frac{0.31}{d}\right)(0.0025)(16°C)^2 \left(\frac{72 \frac{\text{mg}}{\text{L}}}{12 \frac{\text{mg}}{\text{L}} + 72 \frac{\text{mg}}{\text{L}}}\right)$$

$$\times \left(\frac{37 \frac{\text{mg}}{\text{L}}}{0.31 \frac{\text{mg}}{\text{L}} + 37 \frac{\text{mg}}{\text{L}}}\right)$$

$$= 0.17 \text{ d}^{-1}$$

The answer is (C).

Why Other Options Are Wrong

(A) This incorrect solution does not square the temperature term in the corrected maximum growth rate equation. Other definitions and equations are unchanged from the correct solution.

$$\mu_{mD} = kY = \left(\frac{0.38}{d}\right)\left(0.81 \frac{g}{g}\right) = 0.31 \text{ d}^{-1}$$

$$C_N = 29 \frac{\text{mg}}{\text{L}} + 8 \frac{\text{mg}}{\text{L}} = 37 \text{ mg/L}$$

$$\mu'_{mD} = \mu_{mD} 0.0025 T \left(\frac{C_M}{K_M + C_M}\right) \left(\frac{C_N}{K_N + C_N}\right)$$

$$= \left(\frac{0.31}{d}\right)(0.0025)(16°C) \left(\frac{72 \frac{mg}{L}}{12 \frac{mg}{L} + 72 \frac{mg}{L}}\right)$$

$$\times \left(\frac{37 \frac{mg}{L}}{0.31 \frac{mg}{L} + 37 \frac{mg}{L}}\right)$$

$$= 0.011 \text{ d}^{-1}$$

(B) This incorrect solution reverses the values for the half velocity constants. Other definitions and equations are unchanged from the correct solution.

$$\mu_{mD} = kY = \left(\frac{0.38}{d}\right)\left(0.81 \frac{g}{g}\right) = 0.31 \text{ d}^{-1}$$

$$C_N = 29 \frac{mg}{L} + 8 \frac{mg}{L} = 37 \text{ mg/L}$$

$$\mu'_{mD} = \left(\frac{0.31}{d}\right)(0.0025)(16°C)^2 \left(\frac{72 \frac{mg}{L}}{0.31 \frac{mg}{L} + 72 \frac{g}{L}}\right)$$

$$\times \left(\frac{37 \frac{mg}{L}}{12 \frac{mg}{L} + 37 \frac{mg}{L}}\right)$$

$$= 0.15 \text{ d}^{-1}$$

(D) This incorrect solution uses the growth rate instead of the maximum growth rate. Other definitions and equations are unchanged from the correct solution.

$$C_N = 29 \frac{mg}{L} + 8 \frac{mg}{L} = 37 \frac{mg}{L}$$

$$\mu_{mD} = 0.38 \text{ d}^{-1}$$

$$\mu'_{mD} = \left(\frac{0.38}{d}\right)(0.0025)(16°C)^2 \left(\frac{72 \frac{mg}{L}}{12 \frac{mg}{L} + 72 \frac{mg}{L}}\right)$$

$$\times \left(\frac{37 \frac{mg}{L}}{0.31 \frac{mg}{L} + 37 \frac{mg}{L}}\right)$$

$$= 0.21 \text{ d}^{-1}$$

SOLUTION 40

This problem can be solved by directly comparing the biomass produced in each equation. The upper equation is for the anaerobic process and the lower equation is for the aerobic process. This is seen by the occurrence of free oxygen as a reactant in the lower equation. Biomass is represented by $C_5H_7O_2N$. The aerobic process produces 3 moles of biomass for 13 moles of waste and the anaerobic process produces 1 mole of biomass for 18 moles of waste.

y yield   mol/mol

The biomass yield for the aerobic process is

$$y = \frac{n_{biomass}}{n_{waste}} = \frac{3 \text{ mol biomass}}{13 \text{ mol waste}}$$

$$= 0.23 \text{ mol/mol}$$

The biomass yield for the anaerobic process is

$$y = \frac{n_{biomass}}{n_{waste}} = \frac{1 \text{ mol biomass}}{18 \text{ mol waste}}$$

$$= 0.056 \text{ mol/mol}$$

The ratio of aerobic to anaerobic biomass is

$$\frac{0.23 \frac{mol}{mol}}{0.056 \frac{mol}{mol}} = 4:1$$

The answer is (C).

Why Other Options Are Wrong

(A) This incorrect choice uses the direct ratio of anaerobic biomass to aerobic biomass, confusing the two equations. The ratio is reversed because the upper equation represents the anaerobic process and the lower equation represents the aerobic process.

The biomass yield for the anaerobic process is 3 mol.

The biomass yield for the aerobic process is 1 mol.

The ratio of aerobic to anaerobic biomass is 1:3.

(B) This incorrect choice uses the direct ratio of aerobic biomass to anaerobic biomass. The biomass yields shown as a product in each equation are based on different waste quantities shown as reactants in each equation.

The biomass yield for the aerobic process is 3 mol.

The biomass yield for the anaerobic process is 1 mol.

The ratio of aerobic to anaerobic biomass is 3:1.

(D) This incorrect choice uses the ratio of the total moles of waste to the total moles of biomass.

$$\frac{18 \text{ mol waste} + 13 \text{ mol waste}}{1 \text{ mol biomass} + 3 \text{ mol biomass}} = 31:4 \quad (8:1)$$

SOLUTION 41

f	ratio of BOD_5 to BOD_u	
N	effluent ammonia concentration	mg/L
N_o	influent ammonia concentration	mg/L
S	effluent BOD_5	mg/L
S_o	influent BOD_5	mg/L
Q	flow rate	gal/day
X_p	biomass wasted	lbm/day

The daily oxygen requirement must include both oxidation of the BOD and nitrification.

$$\frac{\text{lbm } O_2}{\text{day}} = Q(S_o - S)f - 1.42X_p + 4.57Q(N_o - N)$$

The 1.42 accounts for the theoretical oxygen demand for mineralizing the wasted biomass and the 4.57 accounts for the nitrogenous oxygen demand (4.57 g oxygen per gram of ammonia).

$$\frac{\text{lbm } O_2}{\text{day}} = \left(5.0 \times 10^6 \ \frac{\text{gal}}{\text{day}}\right)\left(3.785 \ \frac{\text{L}}{\text{gal}}\right)$$
$$\times \left(356 \ \frac{\text{mg}}{\text{L}} - 30 \ \frac{\text{mg}}{\text{L}}\right)(1.52)\left(\frac{2.204 \text{ lbm}}{10^6 \text{ mg}}\right)$$
$$- (1.42)\left(2800 \ \frac{\text{lbm}}{\text{day}}\right) + (4.57)$$
$$\times \left(5.0 \times 10^6 \ \frac{\text{gal}}{\text{day}}\right)\left(3.785 \ \frac{\text{L}}{\text{gal}}\right)$$
$$\times \left(\frac{2.204 \text{ lbm}}{10^6 \text{ mg}}\right)\left(63 \ \frac{\text{mg}}{\text{L}} - 10 \ \frac{\text{mg}}{\text{L}}\right)$$
$$= 26{,}795 \text{ lbm/day} \quad (27{,}000 \text{ lbm/day})$$

The answer is (C).

Why Other Options Are Wrong

(A) This incorrect solution fails to include the nitrogenous oxygen demand. Other assumptions, definitions, and equations are the same as used in the correct solution.

$$\text{lbm } O_2/\text{day} = Q(S_o - S)f - 1.42X_p$$
$$= \left(5.0 \times 10^6 \ \frac{\text{gal}}{\text{day}}\right)\left(3.785 \ \frac{\text{L}}{\text{gal}}\right)$$
$$\times \left(356 \ \frac{\text{mg}}{\text{L}} - 30 \ \frac{\text{mg}}{\text{L}}\right)(1.52)$$
$$\times \left(2.204 \ \frac{\text{lbm}}{10^6 \text{ mg}}\right) - (1.42)$$
$$\times \left(2800 \ \frac{\text{lbm}}{\text{day}}\right)$$
$$= 16{,}692 \text{ lbm/day} \quad (17{,}000 \text{ lbm/day})$$

(B) This incorrect solution fails to convert BOD_5 to BOD_u. Other assumptions, definitions, and equations are the same as used in the correct solution.

$$\text{lbm } O_2/\text{day} = Q(S_o - S) - 1.42X_p + 4.57Q$$
$$\times (N_o - N)$$
$$= \left(5.0 \times 10^6 \ \frac{\text{gal}}{\text{day}}\right)\left(3.785 \ \frac{\text{L}}{\text{gal}}\right)$$
$$\times \left(356 \ \frac{\text{mg}}{\text{L}} - 30 \ \frac{\text{mg}}{\text{L}}\right)\left(2.204 \ \frac{\text{lbm}}{10^6 \text{ mg}}\right)$$
$$- (1.42)\left(2800 \ \frac{\text{lbm}}{\text{day}}\right)$$
$$+ (4.57)\left(5.0 \times 10^6 \ \frac{\text{gal}}{\text{day}}\right)\left(3.785 \ \frac{\text{L}}{\text{gal}}\right)$$
$$\times \left(2.204 \ \frac{\text{lbm}}{10^6 \text{ mg}}\right)\left(63 \ \frac{\text{mg}}{\text{L}} - 10 \ \frac{\text{mg}}{\text{L}}\right)$$
$$= 19{,}724 \text{ lbm/day} \quad (20{,}000 \text{ lbm/day})$$

(D) This incorrect solution fails to subtract the theoretical oxygen demand for mineralizing biomass. Other assumptions, definitions, and equations are the same as used in the correct solution.

$$\text{lbm } O_2/\text{day} = Q(S_o - S)f + 4.57Q(N_o - N)$$
$$= \left(5.0 \times 10^6 \ \frac{\text{gal}}{\text{day}}\right)\left(3.785 \ \frac{\text{L}}{\text{gal}}\right)$$
$$\times \left(356 \ \frac{\text{mg}}{\text{L}} - 30 \ \frac{\text{mg}}{\text{L}}\right)(1.52)$$
$$\times \left(2.204 \ \frac{\text{lbm}}{10^6 \text{ mg}}\right)$$
$$+ (4.57)\left(5.0 \times 10^6 \ \frac{\text{gal}}{\text{day}}\right)$$
$$\times \left(3.785 \ \frac{\text{L}}{\text{gal}}\right)\left(2.204 \ \frac{\text{lbm}}{10^6 \text{ mg}}\right)$$
$$\times \left(63 \ \frac{\text{mg}}{\text{L}} - 10 \ \frac{\text{mg}}{\text{L}}\right)$$
$$= 30{,}771 \text{ lbm/day} \quad (31{,}000 \text{ lbm/day})$$

SOLUTION 42

The Clean Water Act (CWA) primarily addresses discharges to waters of the United States by imposing effluent limitations through a pretreatment and permit program. Three major areas of the CWA are discharge criteria, permitting, and priority pollutants.

Discharge criteria provisions set pretreatment standards for categorical discharges from industry to publicly owned treatment works (POTWs). Discharger categories are specifically defined in the regulations and apply to nondomestic waste generators who discharge to a POTW. Discharge criteria also set secondary treatment standards applicable to discharges to receiving waters from POTWs. These include treatment standards for specific "conventional pollutants" such as biochemical oxygen demand (BOD), suspended solids (SS), and pH.

Permitting provisions define the National Pollutant Discharge Elimination System (NPDES) which regulates discharges to waters of the United States. In most cases, the NPDES program is administered by individual states. The NPDES permits consider site-specific conditions in establishing discharge criteria.

The 1977 amendments to CWA included a list of 65 priority pollutants (specific chemicals and classes of chemicals) to be used for defining toxic substances and establishing permit limits. The original list has been expanded to the current list of 129 priority pollutants. Priority pollutants are those chemicals with relatively high toxicity and high production volume.

The Safe Drinking Water Act (SDWA) defines, among other things, maximum contaminant levels (MCLs) for drinking water. The MCLs are frequently used to establish clean-up levels in groundwater contamination remediation programs. The MCLs are not part of the CWA.

The answer is (B).

Why Other Options Are Wrong

(A) This choice is incorrect because categorical pretreatment standards for industrial effluents are included in the Clean Water Act (CWA).

(C) This choice is incorrect because the Clean Water Act (CWA) does include a national permit system for surface water discharges.

(D) This choice is incorrect because the Clean Water Act (CWA) does define specific priority pollutants for regulation by discharge standards.

SOLUTION 43

k_d	endogenous decay coefficient	d^{-1}
K_s	half velocity constant	mg/L
N_o	influent ammonia nitrogen concentration	mg/L
S_o	influent BOD concentration	mg/L
μ_m	maximum growth rate constant	d^{-1}
μ'_m	corrected maximum growth rate constant	d^{-1}
θ_c^m	minimum mean cell residence time	d

For nitrification,

$$\frac{1}{\theta_c^m} = \frac{\mu'_m N_o}{K_s + N_o} - k_d$$

$$= \frac{\left(0.41 \frac{1}{d}\right)\left(51 \frac{mg}{L}\right)}{(2.6 + 51) \frac{mg}{L}} - \frac{0.07}{d}$$

$$= 0.32 \text{ d}^{-1}$$

$$\theta_c^m = 3.1 \text{ d}$$

For BOD removal,

$$\frac{1}{\theta_c^m} = \frac{\mu_m S_o}{K_s + S_o} - k_d$$

$$= \frac{\left(\frac{0.50}{d}\right)\left(312 \frac{mg}{L}\right)}{2.6 \frac{mg}{L} + 312 \frac{mg}{L}} - \frac{0.07}{d}$$

$$= 0.43 \text{ d}^{-1}$$

$$\theta_c^m = 2.3 \text{ d}$$

The greatest minimum mean cell residence controls design. Nitrification controls design with

$$\theta_c^m = 3.1 \text{ d}$$

The answer is (C).

Why Other Options Are Wrong

(A) This incorrect choice calculates the mean cell residence time for biochemical oxygen demand (BOD) removal only. Other definitions and equations are unchanged from the correct solution. For BOD removal,

$$\frac{1}{\theta_c^m} = \frac{\left(\frac{0.50}{d}\right)\left(312 \frac{mg}{L}\right)}{(2.6 + 312) \frac{mg}{L}} - \frac{0.07}{d}$$

$$= 0.43 \text{ d}^{-1}$$

$$\theta_c^m = 2.3 \text{ d}$$

(B) This incorrect choice uses the corrected maximum growth rate constant for calculating the mean cell residence time for both nitrification and for biochemical oxygen demand (BOD) removal and fails to subtract the endogenous decay rate coefficient. Other definitions and equations are unchanged from the correct solution. For nitrification,

$$\frac{1}{\theta_c^m} = \frac{\left(\frac{0.41}{d}\right)\left(51 \frac{mg}{L}\right)}{(2.6 + 51) \frac{mg}{L}}$$

$$= 0.39 \text{ d}^{-1}$$

$$\theta_c^m = 2.6 \text{ d}$$

For BOD removal,

$$\frac{1}{\theta_c^m} = \frac{\left(\frac{0.41}{d}\right)\left(312 \frac{mg}{L}\right)}{(2.6 + 312) \frac{mg}{L}}$$

$$= 0.41 \text{ d}^{-1}$$

$$\theta_c^m = 2.4 \text{ d}$$

Nitrification controls design with

$$\theta_c^m = 2.6 \text{ d}$$

(D) This incorrect choice uses effluent ammonia nitrogen and effluent biochemical oxygen demand (BOD) instead of influent values to calculate the mean cell residence times. Other definitions and equations are unchanged from the correct solution.

| N | effluent ammonia nitrogen concentration | mg/L |
| S | effluent BOD concentration | mg/L |

For nitrification,

$$\frac{1}{\theta_c^m} = \frac{\theta'_m N}{K_s + N} - k_d$$

$$= \frac{\left(\dfrac{0.41}{\text{d}}\right)\left(1 \dfrac{\text{mg}}{\text{L}}\right)}{(2.6 + 1) \dfrac{\text{mg}}{\text{L}}} - \frac{0.07}{\text{d}}$$

$$= 0.044 \text{ d}^{-1}$$

$$\theta_c^m = 23 \text{ d}$$

For BOD removal,

$$\frac{1}{\theta_c^m} = \frac{\mu_m S}{K_s + S} - k_d$$

$$= \frac{\left(\dfrac{0.50}{\text{d}}\right)\left(20 \dfrac{\text{mg}}{\text{L}}\right)}{(2.6 + 20) \dfrac{\text{mg}}{\text{L}}} - \frac{0.07}{\text{d}}$$

$$= 0.37 \text{ d}^{-1}$$

$$\theta_c^m = 2.7 \text{ d}$$

Nitrification controls design with

$$\theta_c^m = 23 \text{ d}$$

AQUATIC BIOLOGY AND MICROBIOLOGY

SOLUTION 44

T_m temperature of the mixed flows °C

$$T_m = \frac{\left(280 \dfrac{\text{m}^3}{\text{s}}\right)(6°\text{C}) + \left(11 \dfrac{\text{m}^3}{\text{s}}\right)(20°\text{C})}{280 \dfrac{\text{m}^3}{\text{s}} + 11 \dfrac{\text{m}^3}{\text{s}}}$$

$$= 6.5°\text{C}$$

| k | river constant | d^{-1} |
| θ | temperature variation constant | – |

For temperatures between 4°C and 20°C, use 1.135 for the temperature variation constant.

$$k_{6.5} = k_6 \theta^{6.5-6} = \left(\frac{0.080}{\text{d}}\right)(1.135^{6.5-6})$$

$$= 0.085 \text{ d}^{-1}$$

The answer is (B).

Why Other Options Are Wrong

(A) This incorrect solution reverses the temperatures in the temperature correction equation. Other assumptions, definitions, and equations are unchanged from the correct solution.

$$T_m = \frac{\left(280 \dfrac{\text{m}^3}{\text{s}}\right)(6°\text{C}) + \left(11 \dfrac{\text{m}^3}{\text{s}}\right)(20°\text{C})}{\left(280 \dfrac{\text{m}^3}{\text{s}} + 11 \dfrac{\text{m}^3}{\text{s}}\right)}$$

$$= 6.5°\text{C}$$

$$k_{6.5} = k_6 \theta^{6-6.5} = \left(\frac{0.080}{\text{d}}\right)(1.135^{6-6.5})$$

$$= 0.075 \text{ d}^{-1}$$

(C) This incorrect solution uses the simple average of the river water and cooling water temperatures. Other assumptions, definitions, and equations are unchanged from the correct solution.

$$T_m = \frac{6°\text{C} + 20°\text{C}}{2} = 13°\text{C}$$

$$k_{13} = k_6 \theta^{13-6} = \left(\frac{0.080}{\text{d}}\right)(1.135^{13-6})$$

$$= 0.19 \text{ d}^{-1}$$

(D) This incorrect solution corrects for the temperature of the cooling water instead of for the mixed flows. Other assumptions, definitions, and equations are unchanged from the correct solution.

$$k_{20} = k_6 \theta^{20-6} = \left(\frac{0.080}{\text{d}}\right)(1.135^{20-6})$$

$$= 0.47 \text{ d}^{-1}$$

SOLUTION 45

Assimilative capacity is the ability of the environment to absorb waste discharges, and is influenced by both stock and fund pollutants. Stock pollutants are substances or materials for which the assimilative capacity is very small—essentially any discharge results in an unacceptable negative impact. Examples of stock pollutants are dioxins and lead, which are compounds that are toxic and that accumulate with little or very slow degradation. Fund pollutants are substances or materials for which the assimilative capacity is relatively large. These are compounds that degrade to produce little accumulation in the environment over time. Examples of fund pollutants are human and animal waste, typically measured by biochemical oxygen demand (BOD) and volatile suspended solids (VSS).

The answer is (D).

Why Other Options Are Wrong

(A) This is an incorrect choice because assimilative capacity is influenced by both fund and stock pollutants.

(B) This is an incorrect choice because assimilative capacity is influenced by both fund and stock pollutants.

(C) This is an incorrect choice because the environment can relatively easily assimilate fund pollutants compared to stock pollutants. Therefore, assimilative capacity is relatively high for fund pollutants and relatively low for stock pollutants.

SOLUTION 46

Assume that the water temperature along the river course is constant and that the water salinity is low (fresh water).

The saturated dissolved oxygen concentration at 8.6°C in fresh water is 11.7 mg/L.

D_o	dissolved oxygen (DO) deficit at discharge point	mg/L
DO_i	DO concentration at discharge point	mg/L
DO_s	saturated DO concentration	mg/L

$$D_o = DO_s - DO_i = 11.7 \, \frac{mg}{L} - 9.3 \, \frac{mg}{L}$$
$$= 2.4 \, mg/L$$

k_d	deoxygenation rate constant	d^{-1}
k_r	reoxygenation rate constant	d^{-1}
L_o	ultimate biochemical oxygen demand (BOD_u) at the discharge point	mg/L
t_c	critical time	d

$$t_c = \left(\frac{1}{k_r - k_d}\right) \ln\left(\left(\frac{k_r}{k_d}\right)\left(1 - \frac{D_o(k_r - k_d)}{k_d L_o}\right)\right)$$

$$= \left(\frac{0.5}{d} - \frac{0.4}{d}\right)^{-1}$$

$$\times \ln\left(\left(\frac{\frac{0.5}{d}}{\frac{0.4}{d}}\right)\left(1 - \frac{\left(2.4 \, \frac{mg}{L}\right)\left(\frac{0.5}{d} - \frac{0.4}{d}\right)}{\left(\frac{0.4}{d}\right)\left(9.8 \, \frac{mg}{L}\right)}\right)\right)$$

$$= 1.6 \, d$$

d_c distance to the monitoring point at t_c mi

$$d_c = (1.6 \, \text{days})\left(0.3 \, \frac{ft}{sec}\right)\left(86{,}400 \, \frac{sec}{day}\right)\left(\frac{1 \, mi}{5280 \, ft}\right)$$
$$= 7.85 \, mi \quad (7.9 \, mi)$$

The answer is (C).

Why Other Options Are Wrong

(A) This incorrect choice fails to invert the rate constant difference in the first term. Other assumptions, definitions, and equations are unchanged from the correct solution.

The saturated dissolved oxygen at 8.6°C in fresh water is 11.7 mg/L.

$$D_o = 11.7 \, \frac{mg}{L} - 9.3 \, \frac{mg}{L}$$
$$= 2.4 \, mg/L$$

$$t_c = (k_r - k_d)\ln\left(\frac{k_r}{k_d}\left(1 - \frac{D_o(k_r - k_d)}{k_d L_o}\right)\right)$$

$$= \left(\frac{0.5}{d} - \frac{0.4}{d}\right)$$

$$\times \ln\left(\left(\frac{\frac{0.5}{d}}{\frac{0.4}{d}}\right)\left(1 - \frac{\left(2.4 \, \frac{mg}{L}\right)\left(\frac{0.5}{d} - \frac{0.4}{d}\right)}{\left(\frac{0.4}{d}\right)\left(9.8 \, \frac{mg}{L}\right)}\right)\right)$$

$$= 0.016 \, d$$

$$d_c = (0.016 \, day)\left(0.3 \, \frac{ft}{sec}\right)\left(86{,}400 \, \frac{sec}{day}\right)\left(\frac{1 \, mi}{5280 \, ft}\right)$$
$$= 0.0785 \, mi \quad (0.079 \, mi)$$

(B) This incorrect choice uses the actual dissolved oxygen concentration at the discharge point instead of the oxygen deficit to calculate critical time. Other assumptions, definitions, and equations are unchanged from the correct solution.

$D_o = 9.3 \text{ mg/L}$

$$t_c = \left(\frac{0.5}{\text{d}} - \frac{0.4}{\text{d}}\right)^{-1}$$

$$\times \ln\left(\left(\frac{\frac{0.5}{\text{d}}}{\frac{0.4}{\text{d}}}\right)\left(1 - \frac{\left(9.3 \frac{\text{mg}}{\text{L}}\right)\left(\frac{0.5}{\text{d}} - \frac{0.4}{\text{d}}\right)}{\left(\frac{0.4}{\text{d}}\right)\left(9.8 \frac{\text{mg}}{\text{L}}\right)}\right)\right)$$

$$= 0.48 \text{ d}$$

The negative sign is ignored.

$$d_c = (0.48 \text{ day})\left(0.3 \frac{\text{ft}}{\text{sec}}\right)\left(86{,}400 \frac{\text{sec}}{\text{day}}\right)\left(\frac{1 \text{ mi}}{5280 \text{ ft}}\right)$$

$$= 2.36 \text{ mi} \quad (2.4 \text{ mi})$$

(D) This incorrect choice confuses dissolved oxygen deficit and ultimate biochemical oxygen demand when calculating the critical time. Other assumptions, definitions, and equations are unchanged from the correct solution.

The saturated dissolved oxygen concentration at 8.6°C in fresh water is 11.7 mg/L.

$$L_o = \text{DO}_s - \text{DO}_i$$
$$= 11.7 \frac{\text{mg}}{\text{L}} - 9.3 \frac{\text{mg}}{\text{L}}$$
$$= 2.4 \text{ mg/L}$$

$$t_c = \left(\frac{0.5}{\text{d}} - \frac{0.4}{\text{d}}\right)^{-1}$$

$$\times \ln\left(\left(\frac{\frac{0.5}{\text{d}}}{\frac{0.4}{\text{d}}}\right)\left(1 - \frac{\left(9.8 \frac{\text{mg}}{\text{L}}\right)\left(\frac{0.5}{\text{d}} - \frac{0.4}{\text{d}}\right)}{\left(\frac{0.4}{\text{d}}\right)\left(2.4 \frac{\text{mg}}{\text{L}}\right)}\right)\right)$$

$$= 36 \text{ d}$$

The negative signs before and after taking the natural log term are ignored.

$$d_c = (36 \text{ days})\left(0.3 \frac{\text{ft}}{\text{sec}}\right)\left(86{,}400 \frac{\text{sec}}{\text{day}}\right)\left(\frac{1 \text{ mi}}{5280 \text{ ft}}\right)$$

$$= 177 \text{ mi} \quad (180 \text{ mi})$$

SOLUTION 47

K_s	half-velocity coefficient	mg/L
S	substrate concentration	mg/L
μ	specific growth rate	d^{-1}
μ_m	maximum specific growth rate	d^{-1}

The Monod kinetic model is expressed as

$$\mu = \mu_m\left(\frac{S}{K_s + S}\right)$$

The half-velocity coefficient is equal to the substrate concentration when the specific growth rate is one-half the maximum growth rate.

$$K_s = S$$
$$\frac{\mu}{\mu_m} = \frac{S}{K_s + S}$$
$$= 0.5$$

$$\mu_m = \frac{\mu}{0.5} = \frac{\frac{6}{\text{d}}}{0.5}$$
$$= 12 \text{ d}^{-1}$$

The answer is (D).

Why Other Options Are Wrong

(A) This incorrect solution confuses the specific growth rate and maximum specific growth rate and assumes that the half-velocity coefficient and substrate concentration are equal. Other assumptions, definitions, and equations are the same as used in the correct solution.

$$\mu = \frac{\left(\frac{6}{\text{d}}\right)\left(16 \frac{\text{mg}}{\text{L}}\right)}{\left(16 \frac{\text{mg}}{\text{L}} + 16 \frac{\text{mg}}{\text{L}}\right)} = 3 \text{ d}^{-1}$$

(B) This incorrect solution assumes that the half-velocity coefficient is equal to half the substrate concentration. Other assumptions, definitions, and equations are the same as used in the correct solution.

$$K_s = 0.5S = (0.5)\left(16 \frac{\text{mg}}{\text{L}}\right)$$
$$= 8 \text{ mg/L}$$

$$\mu = \frac{\left(\frac{6}{\text{d}}\right)\left(16 \frac{\text{mg}}{\text{L}}\right)}{\left(8 \frac{\text{mg}}{\text{L}} + 16 \frac{\text{mg}}{\text{L}}\right)}$$
$$= 4 \text{ d}^{-1}$$

(C) This incorrect solution confuses the specific growth rate and the maximum specific growth rate and assumes that the half-velocity coefficient is equal to one-half the substrate concentration. Other assumptions, definitions, and equations are the same as used in the correct solution.

$$K_s = 0.5S = (0.5)\left(16 \frac{\text{mg}}{\text{L}}\right)$$
$$= 8 \text{ mg/L}$$

$$\mu_m = \frac{\mu(K_s + S)}{S} = \frac{\left(\frac{6}{\text{d}}\right)\left(8 \frac{\text{mg}}{\text{L}} + 16 \frac{\text{mg}}{\text{L}}\right)}{16 \frac{\text{mg}}{\text{L}}}$$
$$= 9 \text{ d}^{-1}$$

SOLUTION 48

Environmental Protection Agency exposure factors for ingestion of soil by children are

| BW | body weight | 15 kg |
| DI | daily intake | 200 mg |

Calculate the average daily dose.

ADD   average daily dose   mg/kg·d
C concentration   ppb or $\mu g/kg$

$$\text{ADD} = \frac{C(\text{DI})}{\text{BW}}$$

$$\text{ADD}_{As} = \frac{\left(1.3 \, \frac{\mu g}{kg}\right)\left(\frac{1 \, kg}{10^9 \, \mu g}\right)\left(200 \, \frac{mg}{d}\right)}{15 \, kg}$$
$$= 0.000\,000\,017\,3 \text{ mg/kg·d}$$

HQ   hazard quotient   –
RfD reference dose   mg/kg·d

$$\text{HQ} = \frac{\text{ADD}}{\text{RfD}}$$

$$\text{HQ}_{As} = \frac{0.000\,000\,017\,3 \, \frac{mg}{kg \cdot d}}{0.0003 \, \frac{mg}{kg \cdot d}} = 0.000\,058$$

$$\text{ADD}_{Cd} = \frac{\left(0.96 \, \frac{\mu g}{kg}\right)\left(\frac{1 \, kg}{10^9 \, \mu g}\right)\left(200 \, \frac{mg}{d}\right)}{15 \, kg}$$
$$= 0.000\,000\,012\,8 \text{ mg/kg·d}$$

$$\text{HQ}_{Cd} = \frac{0.000\,000\,012\,8 \, \frac{mg}{kg \cdot d}}{0.0005 \, \frac{mg}{kg \cdot d}} = 0.000\,026$$

$$\text{ADD}_F = \frac{\left(0.42 \, \frac{\mu g}{kg}\right)\left(\frac{1 \, kg}{10^9 \, \mu g}\right)\left(200 \, \frac{mg}{d}\right)}{15 \, kg}$$
$$= 0.000\,000\,005\,6 \text{ mg/kg·d}$$

$$\text{HQ}_F = \frac{0.000\,000\,005\,6 \, \frac{mg}{kg \cdot d}}{0.0003 \, \frac{mg}{kg \cdot d}} = 0.000\,019$$

HI   hazard index   –

$$\text{HI} = \sum \text{HQ} = 0.000\,058 + 0.000\,026 + 0.000\,019$$
$$= 0.000\,10$$

The answer is (C).

Why Other Options Are Wrong

(A) This incorrect option calculated chronic daily intake, which applies to carcinogens, instead of average daily dose, which applies to noncarcinogens, and uses the reference dose as a potency factor. The result is risk, not hazard index. This required an assumption that the units given for reference dose should be $(mg/kg \cdot d)^{-1}$ instead of $mg/kg \cdot d$. Other assumptions, definitions, and equations are unchanged from the correct solution.

EPA exposure factors for ingestion of soil by children are

EF	exposure frequency	350 d/yr
ED	exposure duration	6 yr
LT	lifetime	70 yr

CDI   chronic daily intake   mg/kg·d

$$\text{CDI} = \frac{C(\text{DI})(\text{EF})(\text{ED})}{(\text{BW})(\text{LT})}$$

$$\text{CDI}_{As} = \frac{\left(1.3 \, \frac{\mu g}{kg}\right)\left(200 \, \frac{mg}{d}\right)\left(350 \, \frac{d}{yr}\right) \times (6 \, yr)\left(\frac{1 \, kg}{10^9 \, \mu g}\right)}{(15 \, kg)(70 \, yr)\left(365 \, \frac{d}{yr}\right)}$$
$$= 1.4 \times 10^{-9} \text{ mg/kg·d}$$

PF   potency factor $(mg/kg \cdot d)^{-1}$

$$\text{risk} = (\text{CDI})(\text{PF})$$

$$\text{risk}_{As} = \frac{\left(1.4 \times 10^{-9} \, \frac{mg}{kg \cdot d}\right)(0.0003)}{\frac{mg}{kg \cdot d}}$$
$$= 4.2 \times 10^{-13} \text{ mg/kg·d}$$

$$\text{CDI}_{Cd} = \frac{\left(0.96 \, \frac{\mu g}{kg}\right)\left(200 \, \frac{mg}{d}\right)\left(350 \, \frac{d}{yr}\right) \times (6 \, yr)\left(\frac{1 \, kg}{10^9 \, \mu g}\right)}{(15 \, kg)(70 \, yr)\left(365 \, \frac{d}{yr}\right)}$$
$$= 1.1 \times 10^{-9} \text{ mg/kg·d}$$

$$\text{risk}_{Cd} = \frac{\left(1.1 \times 10^{-9} \, \frac{mg}{kg \cdot d}\right)(0.0005)}{\frac{mg}{kg \cdot d}}$$
$$= 5.5 \times 10^{-13} \text{ mg/kg·d}$$

$$\text{CDI}_F = \frac{\left(0.42 \, \frac{\mu g}{kg}\right)\left(200 \, \frac{mg}{d}\right)\left(350 \, \frac{d}{yr}\right) \times (6 \, yr)\left(\frac{1 \, kg}{10^9 \, \mu g}\right)}{(15 \, kg)(70 \, yr)\left(365 \, \frac{d}{yr}\right)}$$

$$= 4.6 \times 10^{-10} \, mg/kg \cdot d$$

$$\text{risk}_F = \frac{\left(4.6 \times 10^{-10} \, \frac{mg}{kg \cdot d}\right)(0.0003)}{\frac{mg}{kg \cdot d}}$$

$$= 1.4 \times 10^{-13}$$

$$\text{HI} = \sum \text{risk}$$
$$= (4.2 \times 10^{-13}) + (5.5 \times 10^{-13}) + (1.4 \times 10^{-13})$$
$$= 1.1 \times 10^{-12}$$

(B) This incorrect option calculated chronic daily intake instead of average daily dose. Other assumptions, definitions, and equations are unchanged from the correct solution.

$$\text{CDI}_{As} = \frac{\left(1.3 \, \frac{\mu g}{kg}\right)\left(200 \, \frac{mg}{d}\right)\left(350 \, \frac{d}{yr}\right) \times (6 \, yr)\left(\frac{1 \, kg}{10^9 \, \mu g}\right)}{(15 \, kg)(70 \, yr)\left(365 \, \frac{d}{yr}\right)}$$

$$= 1.4 \times 10^{-9} \, mg/kg \cdot d$$

$$\text{HQ}_{As} = \frac{1.4 \times 10^{-9} \, \frac{mg}{kg \cdot d}}{0.0003 \, \frac{mg}{kg \cdot d}} = 4.7 \times 10^{-6}$$

$$\text{CDI}_{Cd} = \frac{\left(0.96 \, \frac{\mu g}{kg}\right)\left(200 \, \frac{mg}{d}\right)\left(350 \, \frac{d}{yr}\right) \times (6 \, yr)\left(\frac{1 \, kg}{10^9 \, \mu g}\right)}{(15 \, kg)(70 \, yr)\left(365 \, \frac{d}{yr}\right)}$$

$$= 1.1 \times 10^{-9} \, mg/kg \cdot d$$

$$\text{HQ}_{Cd} = \frac{1.1 \times 10^{-9} \, \frac{mg}{kg \cdot d}}{0.0005 \, \frac{mg}{kg \cdot d}} = 2.2 \times 10^{-6}$$

$$\text{CDI}_F = \frac{\left(0.42 \, \frac{\mu g}{kg}\right)\left(200 \, \frac{mg}{d}\right)\left(350 \, \frac{d}{yr}\right) \times (6 \, yr)\left(\frac{1 \, kg}{10^9 \, \mu g}\right)}{(15 \, kg)(70 \, yr)\left(365 \, \frac{d}{yr}\right)}$$

$$= 4.6 \times 10^{-10} \, mg/kg \cdot d$$

$$\text{HQ}_F = \frac{4.6 \times 10^{-10} \, \frac{mg}{kg \cdot d}}{0.0003 \, \frac{mg}{kg \cdot d}} = 1.5 \times 10^{-6}$$

$$\text{HI} = (4.7 \times 10^{-6}) + (2.2 \times 10^{-6}) + (1.5 \times 10^{-6})$$
$$= 8.3 \times 10^{-6}$$

(D) This incorrect option uses parts per billion as a ratio of mass to volume instead of mass to mass and uses 1 L for the daily intake. Other assumptions, definitions, and equations are unchanged from the correct solution.

EPA exposure factors for ingestion of soil by children are

DI   daily intake   1 L

$$\text{ADD}_{As} = \frac{\left(1 \, \frac{L}{d}\right)\left(1.3 \, \frac{\mu g}{L}\right)\left(\frac{1 \, mg}{10^3 \, \mu g}\right)}{15 \, kg}$$

$$= 0.000\,087 \, mg/kg \cdot d$$

$$\text{HQ}_{As} = \frac{0.000\,087 \, \frac{mg}{kg \cdot d}}{0.0003 \, \frac{mg}{kg \cdot d}} = 0.29$$

$$\text{ADD}_{Cd} = \frac{\left(1 \, \frac{L}{d}\right)\left(0.96 \, \frac{\mu g}{L}\right)\left(\frac{1 \, mg}{10^3 \, \mu g}\right)}{15 \, kg}$$

$$= 0.000\,064 \, mg/kg \cdot d$$

$$\text{HQ}_{As} = \frac{0.000\,064 \, \frac{mg}{kg \cdot d}}{0.0005 \, \frac{mg}{kg \cdot d}} = 0.13$$

$$\text{ADD}_F = \frac{\left(1 \, \frac{L}{d}\right)\left(0.42 \, \frac{\mu g}{L}\right)\left(\frac{1 \, mg}{10^3 \, \mu g}\right)}{15 \, kg}$$

$$= 0.000\,028 \, mg/kg \cdot d$$

$$\text{HQ}_{As} = \frac{0.000\,028 \, \frac{mg}{kg \cdot d}}{0.0003 \, \frac{mg}{kg \cdot d}} = 0.093$$

$$\text{HI} = 0.29 + 0.13 + 0.093$$
$$= 0.51$$

SOLUTION 49

Biochemical oxygen demand is the amount of dissolved oxygen (DO) needed to biologically oxidize organic matter. Theoretical oxygen demand is the stoichiometric oxygen required to completely oxidize organic matter. Therefore, as the BOD to ThOD ratio approaches one for a particular waste, the more likely that waste will be biologically degradable.

The answer is (C).

Why Other Options Are Wrong

(A) This choice is incorrect because it is not the value of a waste's BOD that determines biological degradability, but the comparison of the BOD to ThOD.

(B) This choice is incorrect because it is not the value of a waste's BOD that determines biological degradability, but the comparison of the BOD to ThOD.

(D) This choice is incorrect because the smallest BOD to ThOD ratio represents the chemical waste that is most unlikely to be biologically degradable. As the ratio becomes smaller, the difference between the two parameters becomes greater, representing a lesser proportion of biologically degradable matter.

SOLUTION 50

dilution (%)	concentration (%)	survivors at 96-h	mortality at 96-h (%)
98	2	19	5
96	4	12	40
92	8	7	65
84	16	0	100

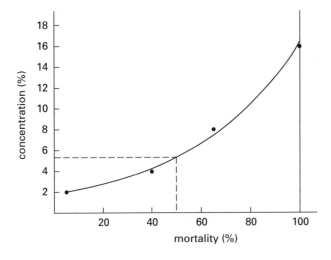

From the illustration, the 96-h LC50 occurs at 5.4% of predilution concentration.

$$\text{96-h LC50} = (\text{predilution concentration})$$
$$\times \left(\frac{\begin{array}{c}\text{\% of predilution concentration} \\ \text{at 50\% mortality}\end{array}}{100\%} \right)$$
$$= \left(1.63 \, \frac{\text{mg}}{\text{L}}\right)\left(\frac{5.4\%}{100\%}\right)$$
$$= 0.088 \text{ mg/L}$$

The answer is (A).

Why Other Options Are Wrong

(B) This incorrect solution uses linear interpolation to find the concentration. The table is unchanged from the correct solution.

$$4\% + \frac{(10-7)(8\% - 4\%)}{12 - 7} = 6.4\%$$
$$\text{96-h LC50} = \left(1.63 \, \frac{\text{mg}}{\text{L}}\right)(0.064)$$
$$= 0.10 \text{ mg/L}$$

(C) This incorrect solution divides the concentration by the 96-h LC50 values in %. The table and illustration are unchanged from the correct solution.

From the illustration, the 96-h LC50 occurs at 5.4%.

$$\text{96-h LC50} = \frac{1.63 \, \frac{\text{mg}}{\text{L}}}{5.4\%}$$
$$= 0.30 \text{ mg/L}$$

(D) This incorrect solution gives the result based on survival instead of mortality. The table and illustration are unchanged from the correct solution.

From the illustration, the 96-h LC50 occurs at 5.4%.

$$\text{96-h LC50} = \left(1.63 \, \frac{\text{mg}}{\text{L}}\right)(1 - 0.054)$$
$$= 1.5 \text{ mg/L}$$

SOLUTION 51

Algal growth is stimulated by excess nutrients, such as nitrogen, that may be discharged with wastewater effluents. Nitrification and denitrification are commonly applied to control excess nutrients in wastewater discharges and reduce their influence on algal growth in receiving waters. Microscreening may be used to physically remove algae as a final step in wastewater treatment, and is especially suitable where treatment involves stabilization ponds. It may also be appropriate in some cases to apply chlorination to deactivate algae and other organisms and to remove excess nutrients by oxidizing ammonia. Aeration would likely have little effect as an algae control strategy, but may be effective for materials other than algae that are often associated with taste and odor.

The answer is (A).

Why Other Options Are Wrong

(B) This is an incorrect choice because microscreening is applied as a physical process for removing algae from stabilization pond and other wastewater treatment process effluents.

(C) This is an incorrect choice because nitrification and denitrification are routinely employed to remove nitrogen from wastewater effluents. Excess nitrogen in wastewater discharges is commonly associated with algal blooms in receiving waters.

(D) This is an incorrect choice because chlorination can be successfully applied to directly deactivate algae as well as to indirectly control algae by oxidizing ammonia. For wastewater effluents, chlorination will usually be followed by dechlorination.

SOLUTION 52

k	growth rate constant	d^{-1}
Y	yield coefficient	g/g
μ_m	maximum growth rate	d^{-1}

$$\mu_m = kY = \left(\frac{2.1}{d}\right)\left(0.23 \frac{g}{g}\right)$$
$$= 0.48 \ d^{-1}$$

DO	dissolved oxygen concentration	mg/L
K_o	half-saturation constant for DO	mg/L
T	temperature	°C
μ'_m	corrected maximum growth rate	d^{-1}

Assume that a typical value for K_o is 1.3 mg/L.

$$\mu'_m = \mu_m e^{(0.098)(T-15)} \left(\frac{DO}{K_o + DO}\right)$$
$$\times (1 - 0.833(7.2 - pH))$$
$$= \left(\frac{0.48}{d}\right)(e^{(0.098)(17-15)}) \left(\frac{7.2 \frac{mg}{L}}{1.3 \frac{mg}{L} + 7.2 \frac{mg}{L}}\right)$$
$$\times (1 - (0.833)(7.2 - 6.4))$$
$$= 0.17 \ d^{-1}$$

The answer is (B).

Why Other Options Are Wrong

(A) This incorrect solution assumes a minimum dissolved oxygen concentration of 2.0 mg/L and uses it in place of the given value. Other assumptions, definitions, and equations are unchanged from the correct solution.

$$\mu_m = \left(\frac{2.1}{d}\right)\left(0.23 \frac{g}{g}\right) = 0.48 \ d^{-1}$$

$$\mu'_m = \left(\frac{0.48}{d}\right)(e^{(0.098)(17-15)}) \left(\frac{2.0 \frac{mg}{L}}{1.3 \frac{mg}{L} + 7.2 \frac{mg}{L}}\right)$$
$$\times (1 - (0.833)(7.2 - 6.4))$$
$$= 0.046 \ d^{-1}$$

(C) This incorrect solution calculates the maximum growth, but does not apply corrections. Other assumptions, definitions, and equations are unchanged from the correct solution.

$$\mu_m = \left(\frac{2.1}{d}\right)\left(0.23 \frac{g}{g}\right) = 0.48 \ d^{-1}$$

(D) This incorrect solution uses the value of the growth rate constant for the maximum growth rate. Other assumptions, definitions, and equations are unchanged from the correct solution.

$$\mu'_m = \left(\frac{2.1}{d}\right)(e^{(0.098)(17-15)}) \left(\frac{7.2 \frac{mg}{L}}{1.3 \frac{mg}{L} + 7.2 \frac{mg}{L}}\right)$$
$$\times (1 - (0.833)(7.2 - 6.4))$$
$$= 0.72 \ d^{-1}$$

SOLUTION 53

Because the lake has a constant volume, the flow rates in and out of the lake are assumed to be equal at 735,000 ft^3/yr.

Assume that all the ammonia depletion in the lake is from nitrification.

$NH_{3(in)}$	ammonia mass in inflow	mg/yr
$NH_{3(out)}$	ammonia mass in outflow	mg/yr
ΔNH_3	ammonia mass lost to nitrification	mg/yr

$$NH_{3(in)} = NH_{3(out)} + \Delta NH_3$$

$$\left(735{,}000 \ \frac{ft^3}{yr}\right)\left(1.2 \ \frac{mg}{L}\right)\left(28.3 \ \frac{L}{ft^3}\right)$$
$$= \left(735{,}000 \ \frac{ft^3}{yr}\right)\left(0.26 \ \frac{mg}{L}\right)\left(28.3 \ \frac{L}{ft^3}\right)$$
$$+ \Delta NH_3$$

$$\Delta NH_3 = \left(1.2 \ \frac{mg}{L} - 0.26 \ \frac{mg}{L}\right)\left(735{,}000 \ \frac{ft^3}{yr}\right)$$
$$\times \left(28.3 \ \frac{L}{ft^3}\right)$$
$$= 2.0 \times 10^7 \ mg/yr$$

Assume that the ammonia concentration is given in the customary form as N (NH_3 as N).

NOD	nitrogenous oxygen demand for nitrification	4.57 g O_2/g N
OD_e	oxygen demand exerted	kg O_2/yr

$$OD_e = (\Delta NH_3)(NOD)$$
$$= \left(2.0 \times 10^7 \; \frac{\text{mg NH}_3 \text{ as N}}{\text{yr}}\right)\left(4.57 \; \frac{\text{g O}_2}{\text{g N}}\right)$$
$$\times \left(\frac{1 \text{ g}}{1000 \text{ mg}}\right)\left(\frac{1 \text{ kg}}{1000 \text{ g}}\right)$$
$$= 91 \text{ kg O}_2/\text{yr}$$

The answer is (C).

Why Other Options Are Wrong

(A) This incorrect choice uses only the outflow ammonia concentration instead of taking the difference between inflow and outflow concentrations. Other assumptions, definitions, and equations are the same as used in the correct solution.

$$\Delta NH_3 = \left(0.26 \; \frac{\text{mg}}{\text{L}}\right)\left(735{,}000 \; \frac{\text{ft}^3}{\text{yr}}\right)\left(28.3 \; \frac{\text{L}}{\text{ft}^3}\right)$$
$$= 5.4 \times 10^6 \text{ mg/yr}$$
$$OD_e = \left(5.4 \times 10^6 \; \frac{\text{mg NH}_3 \text{ as N}}{\text{yr}}\right)\left(4.57 \; \frac{\text{g O}_2}{\text{g N}}\right)$$
$$\times \left(\frac{1 \text{ g}}{1000 \text{ mg}}\right)\left(\frac{1 \text{ kg}}{1000 \text{ g}}\right)$$
$$= 25 \text{ kg O}_2/\text{yr}$$

(B) This incorrect choice uses the mole ratio of oxygen to nitrogen instead of the nitrogenous oxygen demand. Other assumptions, definitions, and equations are the same as used in the correct solution.

$$\Delta NH_3 = \left(1.2 \; \frac{\text{mg}}{\text{L}} - 0.26 \; \frac{\text{mg}}{\text{L}}\right)\left(735{,}000 \; \frac{\text{ft}^3}{\text{yr}}\right)$$
$$\times \left(28.3 \; \frac{\text{L}}{\text{ft}^3}\right)$$
$$= 2.0 \times 10^7 \text{ mg/yr}$$
$$OD_e = \left(2.0 \times 10^7 \; \frac{\text{mg NH}_3 \text{ as N}}{\text{yr}}\right)\left(\frac{32 \text{ mol O}_2}{14 \text{ mol N}}\right)$$
$$\times \left(\frac{1 \text{ g}}{1000 \text{ mg}}\right)\left(\frac{1 \text{ kg}}{1000 \text{ g}}\right)$$
$$= 46 \text{ kg O}_2/\text{yr}$$

(D) This incorrect choice uses the lake volume instead of performing a mass balance. This requires using inappropriate volume/time units for volume. Other assumptions, definitions, and equations are the same as used in the correct solution.

$$\Delta NH_3 = \left(1.2 \; \frac{\text{mg}}{\text{L}} - 0.26 \; \frac{\text{mg}}{\text{L}}\right)\left(1.9 \times 10^6 \; \frac{\text{ft}^3}{\text{yr}}\right)$$
$$\times \left(28.3 \; \frac{\text{L}}{\text{ft}^3}\right)$$
$$= 5.1 \times 10^7 \text{ mg/yr}$$

$$OD_e = \left(4.3 \times 10^7 \; \frac{\text{mg NH}_3 \text{ as N}}{\text{yr}}\right)\left(4.57 \; \frac{\text{g O}_2}{\text{g N}}\right)$$
$$\times \left(\frac{1 \text{ g}}{1000 \text{ mg}}\right)\left(\frac{1 \text{ kg}}{1000 \text{ g}}\right)$$
$$= 233 \text{ kg O}_2/\text{yr} \quad (230 \text{ kg O}_2/\text{yr})$$

SOLUTION 54

Assume that the Thomas equation provides an adequate estimate of the most probable number (MPN).

$$\frac{\text{MPN}}{100 \text{ mL}} = \frac{(100)(\text{positives})}{\sqrt{\text{mL negatives}}\sqrt{\frac{\text{mL positives}}{+ \text{ mL negatives}}}}$$
$$= \frac{(100)(3+2+1)}{\sqrt{\begin{array}{c}(2)(1 \text{ mL}) + (3)(0.1 \text{ mL}) \\ + (4)(0.01 \text{ mL})\end{array}}}$$
$$\times \sqrt{\begin{array}{c}(5)(1 \text{ mL}) + (5)(0.1 \text{ mL}) \\ + (5)(0.01 \text{ mL})\end{array}}$$
$$= 166/100 \text{ mL} \quad (170/100 \text{ mL})$$

The answer is (D).

Why Other Options Are Wrong

(A) This incorrect solution fails to multiply the total number of positives by 100. Assumptions and equations are unchanged from the correct solution.

$$\frac{\text{MPN}}{100 \text{ mL}} = \frac{3+2+1}{\sqrt{\begin{array}{c}(2)(1 \text{ mL}) + (3)(0.1 \text{ mL}) \\ + (4)(0.01 \text{ mL})\end{array}}}$$
$$\times \sqrt{\begin{array}{c}(5)(1 \text{ mL}) + (5)(0.1 \text{ mL}) \\ + (5)(0.01 \text{ mL})\end{array}}$$
$$= 1.66/100 \text{ mL} \quad (2/100 \text{ mL})$$

(B) This incorrect solution includes the results from the 10 mL volume. Assumptions and equations are unchanged from the correct solution.

$$\frac{\text{MPN}}{100 \text{ mL}} = \frac{(100)(5+3+2+1)}{\sqrt{\begin{array}{c}(0)(10 \text{ mL}) + (2)(1 \text{ mL}) + (3)(0.1 \text{ mL}) \\ + (4)(0.01 \text{ mL})\end{array}}}$$
$$\times \sqrt{\begin{array}{c}(5)(10 \text{ mL}) + (5)(1 \text{ mL}) \\ + (5)(0.1 \text{ mL}) + (5)(0.01 \text{ mL})\end{array}}$$
$$= 96/100 \text{ mL} \quad (100/100 \text{ mL})$$

(C) This incorrect solution uses positives in place of negatives in the denominator. Assumptions and equations are unchanged from the correct solution.

$$\frac{\text{MPN}}{100 \text{ mL}} = \frac{(100)(3+2+1)}{\left(\sqrt{\begin{array}{c}(3)(1 \text{ mL}) + (2)(0.1 \text{ mL}) \\ + (1)(0.01 \text{ mL})\end{array}} \times \sqrt{\begin{array}{c}(5)(1 \text{ mL}) + (5)(0.1 \text{ mL}) \\ + (5)(0.01 \text{ mL})\end{array}}\right)}$$

$$= 142/100 \text{ mL} \quad (140/100 \text{ mL})$$

SOLUTION 55

DO_i dissolved oxygen concentration at the discharge   mg/L

$$DO_i = 10.9 \frac{\text{mg}}{\text{L}} - 3.2 \frac{\text{mg}}{\text{L}} = 7.7 \text{ mg/L}$$

By observation, the illustrations for options (B) and (D) do not represent the stream's dissolved oxygen profile below the discharge. The illustration for (B) shows that the dissolved oxygen concentration at the discharge ($t = 0$ d) is not near 7.7 mg/L, as it should be. Also, the illustration for (B) is not consistent with the typical profile for an oxygen sag curve. The illustration for (D) shows the dissolved oxygen concentration after about 11 days to be greater than the saturated dissolved oxygen concentration of 10.9 mg/L. The saturated dissolved oxygen concentration is the maximum value possible in the stream under the conditions given.

D_o dissolved oxygen deficit at the discharge   mg/L
k_d deoxygenation rate constant   d^{-1}
k_r reaeration rate constant   d^{-1}
L_o BOD$_u$ at discharge point   mg/L
t_c time of critical oxygen sag point   d

$$t_c = (k_r - k_d)^{-1} \ln\left(\frac{k_r}{k_d}\right)\left(1 - \frac{D_o(k_r - k_d)}{k_d L_o}\right)$$

$$= \left(\frac{0.07}{\text{d}} - \frac{0.04}{\text{d}}\right)^{-1}$$

$$\times \ln\left(\frac{0.07/\text{d}}{0.04/\text{d}}\left(1 - \frac{\left(3.2 \frac{\text{mg}}{\text{L}}\right)\left(\frac{0.07}{\text{d}} - \frac{0.04}{\text{d}}\right)}{\left(\frac{0.04}{\text{d}}\right)\left(7.2 \frac{\text{mg}}{\text{L}}\right)}\right)\right)$$

$$= 5.1/\text{d}$$

Comparing the illustrations for options (A) and (C) to the critical oxygen sag point, only in the illustration for (A) does the curve match the critical time of 5.1 d. The illustration for (C) is not representative of the stream below the discharge.

The answer is (A).

Why Other Options Are Wrong

(B) This illustration is incorrect because the dissolved oxygen deficit for the illustration does not match the calculated value. From the correct solution, DO_i is 7.7 mg/L. The illustration for option (B) shows that the dissolved oxygen concentration at the discharge ($t = 0$ d) is near 3.2 mg/L, not the correct concentration of 7.7 mg/L. The illustration for (B) is not consistent with the typical profile for an oxygen sag curve.

(C) This illustration is incorrect because the critical time on the curve does not match the calculated value. From the correct solution, t_c is 5.1 d. Comparing the illustration for (C) to the critical oxygen sag point, the curve does not match the critical time of 5.1 d and is not representative of the stream below the discharge.

(D) This illustration is incorrect because it shows the dissolved oxygen concentration after about 11 days to be greater than the saturated dissolved oxygen concentration of 10.9 mg/L. The saturated dissolved oxygen concentration is the maximum value possible in the stream under the conditions given, so the illustration does not represent the stream below the discharge.

SOLUTION 56

f biodegradable fraction of total suspended solids (TSS)   –
G stoichiometric oxygen demand for cell oxidation   1.42 g/g
S_o influent BOD$_5$ mg/L
S_u influent BOD$_u$ mg/L
S_x BOD$_5$ of effluent suspended solids   mg/L
X_e effluent TSS   mg/L

$$S_x = f X_e G \frac{S_o}{S_u}$$

$$= \frac{(0.62)\left(18 \frac{\text{mg}}{\text{L}}\right)\left(1.42 \frac{\text{g}}{\text{g}}\right)\left(217 \frac{\text{mg}}{\text{L}}\right)}{312 \frac{\text{mg}}{\text{L}}}$$

$$= 11 \text{ mg/L}$$

S soluble influent BOD$_5$ escaping treatment   mg/L
S_e effluent BOD$_5$ mg/L

$$S = S_e - S_x = 20 \frac{\text{mg}}{\text{L}} - 11 \frac{\text{mg}}{\text{L}}$$

$$= 9 \text{ mg/L}$$

The answer is (B).

Why Other Options Are Wrong

(A) This incorrect choice does not convert the biochemical oxygen demand associated with the effluent solids from BOD_u to BOD_5. Other definitions and equations are the same as used in the correct solution.

$$S_x = fX_eG = (0.62)\left(18\,\frac{mg}{L}\right)\left(1.42\,\frac{g}{g}\right)$$
$$= 15.8\text{ mg/L}$$
$$S = 20\,\frac{mg}{L} - 15.8\,\frac{mg}{L}$$
$$= 4.2\text{ mg/L}$$

(C) This incorrect choice does not include the stoichiometric oxygen requirement for cell oxidation. Other definitions and equations are the same as used in the correct solution.

$$S_x = fX_e\frac{S_o}{S_u}$$
$$= \frac{(0.62)\left(18\,\frac{mg}{L}\right)\left(217\,\frac{mg}{L}\right)}{312\,\frac{mg}{L}}$$
$$= 7.8\text{ mg/L}$$
$$S = 20\,\frac{mg}{L} - 7.8\,\frac{mg}{L}$$
$$= 12\text{ mg/L}$$

(D) This incorrect choice uses the effluent BOD_5 as the soluble BOD_5 that escaped treatment.

$$S = S_e = 20\text{ mg/L}$$

SOLUTION 57

From the illustration, the river equilibrium dissolved oxygen concentration after full recovery is 10.5 mg/L. At 65% of recovery, the concentration is

$$\left(10.5\,\frac{mg}{L}\right)(0.65) = 6.8\text{ mg/L}$$

From the illustration, a dissolved oxygen concentration of 6.8 mg/L and three days correspond to a mixed-flow BOD_u concentration of 14 mg/L.

Effluent flow is

$$\left(12 \times 10^6\,\frac{gal}{day}\right)\left(0.134\,\frac{ft^3}{gal}\right)\left(\frac{1\text{ day}}{86{,}400\text{ sec}}\right)$$
$$= 18.6\text{ ft}^3/\text{sec}$$

S wastewater BOD_u mg/L

For complete mixing downstream of the discharge,

$$\left(37\,\frac{ft^3}{sec}\right)\left(6\,\frac{mg}{L}\right) + \left(18.6\,\frac{ft^3}{sec}\right)S$$
$$= \left(37\,\frac{ft^3}{sec} + 18.6\,\frac{ft^3}{sec}\right)\left(14\,\frac{mg}{L}\right)$$

$$S = \frac{\left(37\,\frac{ft^3}{sec} + 18.6\,\frac{ft^3}{sec}\right)\left(14\,\frac{mg}{L}\right) - \left(37\,\frac{ft^3}{sec}\right)\left(6\,\frac{mg}{L}\right)}{18.6\,\frac{ft^3}{sec}}$$

$$= 30\text{ mg/L}$$

The answer is (B).

Why Other Options Are Wrong

(A) This incorrect solution uses the mixed-flow BOD_u, taken from the illustration, for the allowable wastewater BOD.

From the illustration, the river equilibrium dissolved oxygen concentration after full recovery is 10.5 mg/L. At 65% of recovery, the concentration is

$$\left(10.5\,\frac{mg}{L}\right)(0.65) = 6.8\text{ mg/L}$$

From the illustration, a dissolved oxygen concentration of 6.8 mg/L and three days correspond to a wastewater BOD_u concentration of 14 mg/L.

(C) This incorrect solution misreads 12 MGD as 12 ft³/sec. Other definitions are unchanged from the correct solution.

From the illustration, the river equilibrium dissolved oxygen concentration after full recovery is 10.5 mg/L. At 65% of recovery, the concentration is

$$\left(10.5\,\frac{mg}{L}\right)(0.65) = 6.8\text{ mg/L}$$

From the illustration, a dissolved oxygen concentration of 6.8 mg/L and three days correspond to a mixed BOD_u concentration of 14 mg/L.

For complete mixing downstream of the discharge,

$$\left(37\,\frac{ft^3}{sec}\right)\left(6\,\frac{mg}{L}\right) + \left(12\,\frac{ft^3}{sec}\right)S$$
$$= \left(37\,\frac{ft^3}{sec} + 12\,\frac{ft^3}{sec}\right)\left(14\,\frac{mg}{L}\right)$$

$$S = \cfrac{\left(37 \, \cfrac{\text{ft}^3}{\text{sec}} + 12 \, \cfrac{\text{ft}^3}{\text{sec}}\right)\left(14 \, \cfrac{\text{mg}}{\text{L}}\right) - \left(37 \, \cfrac{\text{ft}^3}{\text{sec}}\right)\left(6 \, \cfrac{\text{mg}}{\text{L}}\right)}{12 \, \cfrac{\text{ft}^3}{\text{sec}}}$$

$$= 39 \text{ mg/L}$$

(D) This incorrect solution uses the recovered dissolved oxygen concentration as the dissolved oxygen concentration at the discharge. Other definitions are unchanged from the correct solution.

From the illustration, the river equilibrium dissolved oxygen concentration after full recovery is 7.3 mg/L. At 65% of recovery, the concentration is

$$\left(7.3 \, \frac{\text{mg}}{\text{L}}\right)(0.65) = 4.7 \text{ mg/L}$$

From the illustration, a dissolved oxygen concentration of 4.7 mg/L and three days correspond to a mixed BOD_u concentration of 22 mg/L.

Effluent flow is

$$\left(12 \times 10^6 \, \frac{\text{gal}}{\text{day}}\right)\left(0.134 \, \frac{\text{ft}^3}{\text{gal}}\right)\left(\frac{1 \text{ day}}{86{,}400 \text{ sec}}\right)$$

$$= 18.6 \text{ ft}^3/\text{sec}$$

For complete mixing downstream of the discharge,

$$\left(37 \, \frac{\text{ft}^3}{\text{sec}}\right)\left(6 \, \frac{\text{mg}}{\text{L}}\right) + \left(18.6 \, \frac{\text{ft}^3}{\text{sec}}\right) S$$

$$= \left(37 \, \frac{\text{ft}^3}{\text{sec}} + 18.6 \, \frac{\text{ft}^3}{\text{sec}}\right)\left(22 \, \frac{\text{mg}}{\text{L}}\right)$$

$$S = \cfrac{\left(37 \, \cfrac{\text{ft}^3}{\text{sec}} + 18.6 \, \cfrac{\text{ft}^3}{\text{sec}}\right)\left(22 \, \cfrac{\text{mg}}{\text{L}}\right) - \left(37 \, \cfrac{\text{ft}^3}{\text{sec}}\right)\left(6 \, \cfrac{\text{mg}}{\text{L}}\right)}{18.6 \, \cfrac{\text{ft}^3}{\text{sec}}}$$

$$= 54 \text{ mg/L}$$

SOLUTION 58

Using the chemical formula for biomass, for every 5 mol of influent carbon, 1 mol of nitrogen is required. The chemical formula, molecular weight, and concentration of acetic acid determine the moles of influent carbon.

The molecular formula for acetic acid is CH_3COOH.

The mole weight of acetic acid is

$$(2)\left(12 \, \frac{\text{g}}{\text{mol}}\right) + (4)\left(1 \, \frac{\text{g}}{\text{mol}}\right) + (2)\left(16 \, \frac{\text{g}}{\text{mol}}\right)$$

$$= 60 \text{ g/mol}$$

The daily molar quantity of acetic acid in the wastewater is

$$\cfrac{\left(310 \, \cfrac{\text{mg}}{\text{L}}\right)\left(\cfrac{1 \text{ g}}{1000 \text{ mg}}\right)\left(12\,000 \, \cfrac{\text{m}^3}{\text{d}}\right)}{\left(60 \, \cfrac{\text{g}}{\text{mol}}\right)\left(\cfrac{1 \text{ m}^3}{1000 \text{ L}}\right)} = 62\,000 \text{ mol/d}$$

The ratio of nitrogen to carbon in the biomass chemical formula is used to find the moles of nitrogen required.

$$\frac{1 \text{ mol N}}{5 \text{ mol C}} = \cfrac{\cfrac{\text{mol N}}{\text{d}}}{62\,000 \, \cfrac{\text{mol C}}{\text{d}}}$$

$$\frac{\text{mol N}}{\text{d}} = \cfrac{\left(62\,000 \, \cfrac{\text{mol C}}{\text{d}}\right)(1 \text{ mol N})}{5 \text{ mol C}}$$

$$= 12\,400 \text{ mol/d}$$

Multiplying the moles of nitrogen required per day by the molecular weight of nitrogen gives the daily mass of nitrogen required.

$$\left(12\,400 \, \frac{\text{mol}}{\text{d}}\right)\left(14 \, \frac{\text{g}}{\text{mol}}\right)\left(\frac{1 \text{ kg}}{1000 \text{ g}}\right)$$

$$= 174 \text{ kg/d} \quad (170 \text{ kg/d})$$

The answer is (B).

Why Other Options Are Wrong

(A) This incorrect solution confuses the molecular weights of carbon and nitrogen.

The mole weight of acetic acid is

$$(2)\left(14 \, \frac{\text{g}}{\text{mol}}\right) + (4)\left(1 \, \frac{\text{g}}{\text{mol}}\right) + (2)\left(16 \, \frac{\text{g}}{\text{mol}}\right) = 64 \text{ g/mol}$$

The daily molar quantity of acetic acid in the wastewater is

$$\cfrac{\left(310 \, \cfrac{\text{mg}}{\text{L}}\right)\left(\cfrac{1 \text{ g}}{1000 \text{ mg}}\right)\left(12\,000 \, \cfrac{\text{m}^3}{\text{d}}\right)}{\left(64 \, \cfrac{\text{g}}{\text{mol}}\right)\left(\cfrac{1 \text{ m}^3}{1000 \text{ L}}\right)} = 58\,125 \text{ mol/d}$$

The ratio of nitrogen to carbon in the biomass chemical formula is used to find the moles of nitrogen required.

$$\frac{1 \text{ mol N}}{5 \text{ mol C}} = \frac{\frac{\text{mol N}}{\text{d}}}{58\,125 \; \frac{\text{mol C}}{\text{d}}}$$

$$\frac{\text{mol N}}{\text{d}} = \frac{\left(58\,125 \; \frac{\text{mol C}}{\text{d}}\right)(1 \text{ mol N})}{5 \text{ mol C}}$$

$$= 11\,625 \text{ mol/d}$$

Multiplying the moles of nitrogen required per day by the molecular weight of nitrogen gives the daily mass of nitrogen required.

$$\left(11\,625 \; \frac{\text{mol}}{\text{d}}\right)\left(12 \; \frac{\text{g}}{\text{mol}}\right)\left(\frac{1 \text{ kg}}{1000 \text{ g}}\right) = 140 \text{ kg/d}$$

(C) This incorrect solution uses the wrong chemical formula for acetic acid.

The molecular formula for acetic acid is CHOOH.

The mole weight of acetic acid is

$$(1)\left(12 \; \frac{\text{g}}{\text{mol}}\right) + (2)\left(1 \; \frac{\text{g}}{\text{mol}}\right) + (2)\left(16 \; \frac{\text{g}}{\text{mol}}\right)$$

$$= 46 \text{ g/mol}$$

The daily molar quantity of acetic acid in the wastewater is

$$\frac{\left(310 \; \frac{\text{mg}}{\text{L}}\right)\left(\frac{1 \text{ g}}{1000 \text{ mg}}\right)\left(12\,000 \; \frac{\text{m}^3}{\text{d}}\right)}{\left(46 \; \frac{\text{g}}{\text{mol}}\right)\left(\frac{1 \text{ m}^3}{1000 \text{ L}}\right)} = 80\,870 \text{ mol/d}$$

The ratio of nitrogen to carbon in the biomass chemical formula is used to find the moles of nitrogen required.

$$\frac{1 \text{ mol N}}{5 \text{ mol C}} = \frac{\frac{\text{mol N}}{\text{d}}}{80\,870 \; \frac{\text{mol C}}{\text{d}}}$$

$$\frac{\text{mol N}}{\text{d}} = \frac{\left(80\,870 \; \frac{\text{mol C}}{\text{d}}\right)(1 \text{ mol N})}{5 \text{ mol C}}$$

$$= 16\,174 \text{ mol/d}$$

Multiplying the moles of nitrogen required per day by the molecular weight of nitrogen gives the daily mass of nitrogen required.

$$\left(16\,174 \; \frac{\text{mol}}{\text{d}}\right)\left(14 \; \frac{\text{g}}{\text{mol}}\right)\left(\frac{1 \text{ kg}}{1000 \text{ g}}\right)$$

$$= 226 \text{ kg/d} \quad (230 \text{ kg/d})$$

(D) This incorrect solution uses the ratio of the biomass nitrogen to acetic acid carbon instead of the biomass nitrogen to biomass carbon.

The mole weight of acetic acid is

$$(2)\left(12 \; \frac{\text{g}}{\text{mol}}\right) + (4)\left(1 \; \frac{\text{g}}{\text{mol}}\right) + (2)\left(16 \; \frac{\text{g}}{\text{mol}}\right)$$

$$= 60 \text{ g/mol}$$

The daily molar quantity of acetic acid in the wastewater is

$$\frac{\left(310 \; \frac{\text{mg}}{\text{L}}\right)\left(\frac{1 \text{ g}}{1000 \text{ mg}}\right)\left(12\,000 \; \frac{\text{m}^3}{\text{d}}\right)}{\left(60 \; \frac{\text{g}}{\text{mol}}\right)\left(\frac{1 \text{ m}^3}{1000 \text{ L}}\right)}$$

$$= 62\,000 \text{ mol/d}$$

Using the chemical formula for acetic acid and biomass, for every 2 mol of influent carbon from acetic acid, 1 mol of nitrogen is produced in the biomass.

$$\frac{1 \text{ mol N}}{2 \text{ mol C}} = \frac{\frac{\text{mol N}}{\text{d}}}{62\,000 \; \frac{\text{mol C}}{\text{d}}}$$

$$\frac{\text{mol N}}{\text{d}} = \frac{\left(62\,000 \; \frac{\text{mol C}}{\text{d}}\right)(1 \text{ mol N})}{2 \text{ mol C}}$$

$$= 31\,000 \text{ mol/d}$$

Multiplying the moles of nitrogen required per day by the molecular weight of nitrogen gives the daily mass of nitrogen required.

$$\left(31\,000 \; \frac{\text{mol}}{\text{d}}\right)\left(14 \; \frac{\text{g}}{\text{mol}}\right)\left(\frac{1 \text{ kg}}{1000 \text{ g}}\right)$$

$$= 434 \text{ kg/d} \quad (430 \text{ kg/d})$$

SOLUTION 59

The design criteria are based on population equivalents (PE). A PE represents the equivalent organic loading from one person. Assume each person in the community contributes one PE.

A surface area m^2

$$A = (800 \text{ people})\left(1 \; \frac{\text{PE}}{\text{person}}\right)\left(6.0 \; \frac{\text{m}^2}{\text{PE}}\right)$$

$$= 4800 \text{ m}^2$$

d bed depth   m
V empty-bed volume m^3

$$V = Ad = (4800 \text{ m}^2)(0.75 \text{ m})$$
$$= 3600 \text{ m}^3$$

Q flow rate m^3/d

$$Q = (800 \text{ people})\left(1 \frac{\text{PE}}{\text{person}}\right)\left(0.2 \frac{\text{m}^3}{\text{PE·d}}\right)$$
$$= 160 \text{ m}^3/\text{d}$$

t empty-bed hydraulic residence time d

$$t = \frac{V}{Q} = \frac{3600 \text{ m}^3}{160 \frac{\text{m}^3}{\text{d}}}$$
$$= 22.5 \text{ d} \quad (23 \text{ d})$$

The answer is (C).

Why Other Options Are Wrong

(A) This incorrect calculation results from confusion about what to do with the population equivalent (PE). Other definitions and equations are the same as used in the correct solution.

$$t = \frac{\left(6.0 \frac{\text{m}^2}{\text{PE}}\right)(0.75 \text{ m})}{\left(0.2 \frac{\text{m}^3}{\text{PE·d}}\right)(800 \text{ people})} = 0.028 \text{ d}$$

(B) This incorrect calculation misuses the relationship among flow, volume, and hydraulic residence time. Other definitions and equations are the same as used in the correct solution.

$$A = (800 \text{ people})\left(1 \frac{\text{PE}}{\text{person}}\right)\left(6.0 \frac{\text{m}^2}{\text{PE}}\right)$$
$$= 4800 \text{ m}^2$$
$$V = (4800 \text{ m}^2)(0.75 \text{ m}) = 3600 \text{ m}^3$$
$$Q = (800 \text{ people})\left(1 \frac{\text{PE}}{\text{person}}\right)\left(0.2 \frac{\text{m}^3}{\text{PE·d}}\right)$$
$$= 160 \text{ m}^3/\text{d}$$
$$t = \frac{Q}{V} = \frac{160 \frac{\text{m}^3}{\text{d}}}{3600 \text{ m}^3}$$
$$= 0.044 \text{ d}$$

Units do not work.

(D) This incorrect calculation divides the area by the depth instead of multiplying to get volume. Other definitions and equations are the same as used in the correct solution.

$$A = (800 \text{ people})\left(1 \frac{\text{PE}}{\text{person}}\right)\left(6.0 \frac{\text{m}^2}{\text{PE}}\right)$$
$$= 4800 \text{ m}^2$$

$$V = \frac{A}{d} = \frac{4800 \text{ m}^2}{0.75 \text{ m}} = 6400 \text{ m}^3$$

Units do not work.

$$Q = (800 \text{ people})\left(1 \frac{\text{PE}}{\text{person}}\right)\left(0.2 \frac{\text{m}^3}{\text{PE·d}}\right)$$
$$= 160 \frac{\text{m}^3}{\text{d}}$$

t empty bed hydraulic residence time d

$$t = \frac{6400 \text{ m}^3}{160 \frac{\text{m}^3}{\text{d}}} = 40 \text{ d}$$

SOLUTION 60

m	mass of chemical released	lbm
SG	chemical specific gravity	–
V	release volume	ft^3
ρ_c	chemical density	–
ρ_w	water density	lbm/ft^3

$$m = \rho_c V = \text{SG}\rho_w V$$
$$= (0.9)\left(62.4 \frac{\text{lbm}}{\text{ft}^3}\right)(200 \text{ gal})\left(0.134 \frac{\text{ft}^3}{\text{gal}}\right)$$
$$= 1505 \text{ lbm}$$

Assume the ditch cross section is rectangular along its entire length.

A	ditch wetted cross-sectional area	ft^2
d	water depth	ft
w	ditch width	ft

$$A = wd = (6 \text{ ft})(2 \text{ ft})$$
$$= 12 \text{ ft}^2$$

g	gravitational constant	32.2 ft/sec^2
s	ditch slope	ft/ft
v*	shear velocity	ft/day

$$\text{v}^* = \sqrt{gds} = \sqrt{\left(32.2 \frac{\text{ft}}{\text{sec}^2}\right)(2 \text{ ft})\left(\frac{1 \text{ ft}}{100 \text{ ft}}\right)}$$
$$= 0.80 \text{ ft/sec}$$

E_L	longitudinal dispersion coefficient	ft^2/day
v	ditch flow average velocity	ft/day

$$E_L = \frac{0.011 \text{v}^2 w^2}{d\text{v}^*}$$
$$= \frac{(0.011)\left(3 \frac{\text{ft}}{\text{sec}}\right)^2 (6 \text{ ft})^2 \left(86{,}400 \frac{\text{sec}}{\text{day}}\right)}{(2 \text{ ft})\left(0.80 \frac{\text{ft}}{\text{sec}}\right)}$$
$$= 192{,}456 \text{ ft}^2/\text{day}$$

t	time	day
x	downstream distance of travel	ft

$$t = \frac{x}{\text{v}} = \frac{(1.8 \text{ mi})\left(5280 \dfrac{\text{ft}}{\text{mi}}\right)}{\left(3 \dfrac{\text{ft}}{\text{sec}}\right)\left(86{,}400 \dfrac{\text{sec}}{\text{day}}\right)} = 0.037 \text{ day}$$

$C_{\max}$	maximum concentration	mg/L
K	ditch dispersion coefficient	day^{-1}

$$C_{\max} = \frac{Me^{-Kt}}{A\sqrt{4\pi E_L t}}$$

$$= \frac{(1505 \text{ lbm})\left(e^{(-0.3/\text{day})(0.037 \text{ day})}\right) \times \left(\dfrac{10^6 \text{ mg}}{2.204 \text{ lbm}}\right)\left(0.0354 \dfrac{\text{ft}^3}{\text{L}}\right)}{(12 \text{ ft}^2)\sqrt{4\pi\left(192{,}456 \dfrac{\text{ft}^2}{\text{day}}\right)(0.037 \text{ day})}}$$

$$= 6660 \text{ mg/L} \quad (6700 \text{ mg/L})$$

The answer is (B).

Why Other Options Are Wrong

(A) This incorrect solution neglects to take the square root of shear velocity. Other definitions, assumptions, and equations are the same as used in the correct solution.

$$M = (0.9)(200 \text{ gal})\left(0.134 \dfrac{\text{ft}^3}{\text{gal}}\right)\left(62.4 \dfrac{\text{lbm}}{\text{ft}^3}\right)$$
$$= 1505 \text{ lbm}$$
$$A = (6 \text{ ft})(2 \text{ ft}) = 12 \text{ ft}^2$$
$$\text{v}^* = gds = \left(32.2 \dfrac{\text{ft}}{\text{sec}^2}\right)(2 \text{ ft})\left(\dfrac{1 \text{ ft}}{100 \text{ ft}}\right)$$
$$= 0.64 \text{ ft/sec}$$

Units do not work.

$$E_L = \frac{(0.011)\left(3 \dfrac{\text{ft}}{\text{sec}}\right)^2 (6 \text{ ft})^2 \left(86{,}400 \dfrac{\text{sec}}{\text{day}}\right)}{(2 \text{ ft})\left(0.64 \dfrac{\text{ft}}{\text{sec}}\right)}$$

$$= 240{,}570 \text{ ft}^2/\text{day}$$

$$t = \frac{x}{\text{v}} = \frac{(1.8 \text{ mi})\left(5280 \dfrac{\text{ft}}{\text{mi}}\right)}{\left(3 \dfrac{\text{ft}}{\text{sec}}\right)\left(86{,}400 \dfrac{\text{sec}}{\text{day}}\right)}$$
$$= 0.037 \text{ day}$$

$$C_{\max} = \frac{(1505 \text{ lbm})\left(e^{(-0.3/\text{day})(0.037 \text{ day})}\right) \times \left(\dfrac{10^6 \text{ mg}}{2.204 \text{ lbm}}\right)\left(0.0354 \dfrac{\text{ft}^3}{\text{L}}\right)}{(12 \text{ ft}^2)\sqrt{4\pi\left(240{,}570 \dfrac{\text{ft}^2}{\text{day}}\right)(0.037 \text{ day})}}$$

$$= 5957 \text{ mg/L} \quad (6000 \text{ mg/L})$$

(C) This incorrect solution neglects to take the negative power of e in the maximum concentration equation. Other definitions, assumptions, and equations are the same as used in the correct solution.

$$M = (0.9)(200 \text{ gal})\left(0.134 \dfrac{\text{ft}^3}{\text{gal}}\right)\left(62.4 \dfrac{\text{lbm}}{\text{ft}^3}\right)$$
$$= 1505 \text{ lbm}$$
$$A = (6 \text{ ft})(2 \text{ ft}) = 12 \text{ ft}^2$$
$$\text{v}^* = \sqrt{\left(32.2 \dfrac{\text{ft}}{\text{sec}^2}\right)(2 \text{ ft})\left(\dfrac{1 \text{ ft}}{100 \text{ ft}}\right)}$$
$$= 0.80 \text{ ft/s}$$

$$E_L = \frac{(0.011)\left(3 \dfrac{\text{ft}}{\text{sec}}\right)^2 (6 \text{ ft})^2 \left(86{,}400 \dfrac{\text{sec}}{\text{day}}\right)}{(2 \text{ ft})\left(0.80 \dfrac{\text{ft}}{\text{sec}}\right)}$$

$$= 192{,}456 \text{ ft}^2/\text{day}$$

$$t = \frac{(1.8 \text{ mi})\left(5280 \dfrac{\text{ft}}{\text{mi}}\right)}{\left(3 \dfrac{\text{ft}}{\text{sec}}\right)\left(86{,}400 \dfrac{\text{sec}}{\text{day}}\right)}$$
$$= 0.037 \text{ day}$$

$$C_{\max} = \frac{Me^{-Kt}}{A\sqrt{4\pi E_L t}}$$

$$= \frac{(1505 \text{ lbm})\left(e^{(0.3/\text{day})(0.037 \text{ day})}\right) \times \left(\dfrac{10^6 \text{ mg}}{2.204 \text{ lbm}}\right)\left(0.0354 \dfrac{\text{ft}^3}{\text{L}}\right)}{(12 \text{ ft}^2)\sqrt{4\pi\left(192{,}456 \dfrac{\text{ft}^2}{\text{day}}\right)(0.037 \text{ day})}}$$

$$= 6809 \text{ mg/L} \quad (6800 \text{ mg/L})$$

(D) This incorrect solution uses the conversion for kilometers to feet instead of miles to feet in the time equation. Other definitions, assumptions, and equations are the same as used in the correct solution.

$$M = (0.9)(200 \text{ gal})\left(0.134 \dfrac{\text{ft}^3}{\text{gal}}\right)\left(62.4 \dfrac{\text{lbm}}{\text{ft}^3}\right)$$
$$= 1505 \text{ lbm}$$
$$A = (6 \text{ ft})(2 \text{ ft}) = 12 \text{ ft}^2$$
$$\text{v}^* = \sqrt{\left(32.2 \dfrac{\text{ft}}{\text{sec}^2}\right)(2 \text{ ft})\left(\dfrac{1 \text{ ft}}{100 \text{ ft}}\right)} = 0.80 \text{ ft/s}$$

$$E_L = \frac{(0.011)\left(3\ \frac{\text{ft}}{\text{sec}}\right)^2 (6\ \text{ft})^2 \left(86{,}400\ \frac{\text{sec}}{\text{day}}\right)}{(2\ \text{ft})\left(0.80\ \frac{\text{ft}}{\text{sec}}\right)}$$

$$= 192{,}456\ \text{ft}^2/\text{day}$$

$$t = \frac{(1.8\ \text{mi})\left(3280\ \frac{\text{ft}}{\text{mi}}\right)}{\left(3\ \frac{\text{ft}}{\text{sec}}\right)\left(86{,}400\ \frac{\text{sec}}{\text{day}}\right)}$$

$$= 0.023\ \text{day}$$

$$C_{\max} = \frac{(1505\ \text{lbm})\left(e^{(-0.3/\text{day})(0.023\ \text{day})}\right) \times \left(\frac{10^6\ \text{mg}}{2.204\ \text{lbm}}\right)\left(0.0354\ \frac{\text{ft}^3}{\text{L}}\right)}{(12\ \text{ft}^2)\sqrt{4\pi\left(192{,}456\ \frac{\text{ft}^2}{\text{day}}\right)(0.023\ \text{day})}}$$

$$= 8482\ \text{mg/L}\quad (8500\ \text{mg/L})$$

SOLUTION 61

Bacterial die-off is known to follow first-order kinetics. A representative first-order rate coefficient for bacterial die-off is $1.0\ \text{d}^{-1}$.

C	concentration at time of interest	%
C_o	initial concentration	%
k	first-order rate coefficient	d^{-1}
t	time	d

$$t = \frac{-\ln\dfrac{C}{C_o}}{k} = \frac{-\ln\left(\dfrac{100\% - 90\%}{100\%}\right)}{\dfrac{1.0}{\text{d}}} = 2.3\ \text{d}$$

The answer is (C).

Why Other Options Are Wrong

(A) This incorrect solution calculates the time for 10% die-off instead of 90%. Other assumptions and definitions are the same as used in the correct solution.

$$t = \frac{-\ln\dfrac{90\%}{100\%}}{\dfrac{1.0}{\text{d}}} = 0.11\ \text{d}$$

(B) This incorrect solution calculates the half-life for the bacteria die-off. Other assumptions and definitions are the same as used in the correct solution.

$t_{1/2}$	half-life	d

$$t_{1/2} = \frac{0.693}{k} = \frac{0.693}{\dfrac{1.0}{\text{d}}} = 0.69\ \text{d}$$

(D) This incorrect solution improperly assumes that the first-order rate coefficient for oxygenation or deoxygenation could be used for bacterial die-off. Other assumptions and definitions are the same as used in the correct solution.

A typical first-order rate coefficient for oxygenation/deoxygenation is $0.23\ \text{d}^{-1}$.

$$t = \frac{-\ln\left(\dfrac{100\% - 90\%}{100\%}\right)}{\dfrac{0.23}{\text{d}}} = 10\ \text{d}$$

SOLID AND HAZARDOUS WASTE

SOLUTION 62

A solid waste is a hazardous waste if it exhibits either ignitable, toxic, corrosive, or reactive characteristics as defined by Title 40 of the Code of Federal Regulations (40 CFR), Secs. 261.20 to 261.24, or is listed in 40 CFR, Part 261, Subpart D, Secs. 261.30 to 261.33.

The general criteria for hazardous waste are

characteristic of ignitability (40 CFR, Sec. 261.21)	flash point less than 60°C or 140°F
characteristic of corrosivity (40 CFR, Sec. 261.22)	$2 \geq \text{pH} \geq 12.5$
characteristic of reactivity (40 CFR, Sec. 261.23)	reacts violently with water
characteristic of toxicity (40 CFR, Sec. 261.24)	listed in Table 1 of 40 CFR, Sec. 261.24

Assume that the waste is not listed in Subpart D because no information is given regarding the listing.

The waste meets the criteria for ignitability. This is sufficient by itself to classify the waste as hazardous waste.

The answer is (B).

Why Other Options Are Wrong

(A) This choice is incorrect because the toxaphene concentration by the toxicity characteristic leaching procedure must exceed $0.5\ \text{mg/L}$ for classification as hazardous waste. In mg/L units, the toxaphene concentration in the waste is only $0.0058\ \text{mg/L}$.

(C) This choice is incorrect because hazardous waste classification results when any single characteristic meets the regulatory criteria. Although the waste does not meet the toxicity, corrosivity, or reactivity criteria, it does satisfy the ignitability criteria and is, therefore, hazardous waste.

(D) This choice is incorrect because the waste does meet the ignitability criteria for classification as hazardous waste.

SOLUTION 63

Assume a waste sample of 100 lbm so that discarded mass can be expressed in lbm.

component	discarded mass, $m_{discarded}$ (lbm)	discarded moisture (%)	dry mass, m_{dry} (lbm)	dry density, $\rho_{dry,discarded}$ (lbm/ft³)	volume at 40% moisture, $V_{40\%,moist}$ (ft³)
garden	25	60	10	7	2.4
food	9	70	2.7	18	0.25
paper	38	6	36	5	12

The dry mass is

$$m_{dry} = m_{discarded}\left(\frac{1 - \text{discarded moisture in \%}}{100\%}\right)$$

The volume at 40% moisture is

$$V_{40\%,moist} = \frac{m_{dry}}{\rho_{dry,discarded}(1-0.4)}$$

The paper volume is

$$V_{paper} = (2.4 \text{ ft}^3 + 0.25 \text{ ft}^3)(0.1) = 0.27 \text{ ft}^3 < 12 \text{ ft}^3$$

Paper is not limiting.

The mulch volume per 100 lbm waste is

$$V_{mulch} = 2.4 \text{ ft}^3 + 0.25 \text{ ft}^3 + 0.27 \text{ ft}^3 = 2.9 \text{ ft}^3$$

The potential revenue is

$$\text{revenue} = (100{,}000 \text{ people})\left(5 \frac{\text{lbm waste}}{\text{person-day}}\right)$$
$$\times \left(\frac{2.9 \text{ ft}^3 \text{ mulch}}{100 \text{ lbm waste}}\right)\left(\frac{1 \text{ yd}^3}{27 \text{ ft}^3}\right)$$
$$\times \left(365 \frac{\text{day}}{\text{yr}}\right)\left(\frac{\$10}{1 \text{ yd}^3}\right)$$
$$= \$1.96 \times 10^6/\text{yr} \quad (\$2.0 \text{ million/yr})$$

The answer is (A).

Why Other Options Are Wrong

(B) This incorrect solution bases the paper fraction on mass instead of volume and uses mass weighted average density instead of individual component density. The solution also uses the fraction of moisture instead of the fraction of waste when calculating mulch volume at 40% moisture. Other assumptions are unchanged from the correct solution.

component	discarded mass, $m_{discarded}$ (lbm)	discarded moisture (%)	dry mass, m_{dry} (lbm)	dry density, $\rho_{dry,discarded}$ (lbm/ft³)	volume at 40% moisture, $V_{40\%,moist}$ (ft³)
garden	25	60	10	7	2.4
food	9	70	2.7	18	0.25
paper	38	6	36	5	12

The paper mass in the mulch is

$$m_{paper} = (10 \text{ lbm} + 2.7 \text{ lbm})(0.1) = 1.3 \text{ lbm}$$

The total mass is

$$m_{total} = 10 \text{ lbm} + 2.7 \text{ lbm} + 1.3 \text{ lbm} = 14 \text{ lbm}$$

The average density is

$$\frac{(10 \text{ lbm})\left(7 \frac{\text{lbm}}{\text{ft}^3}\right) + (2.7 \text{ lbm}) \times \left(18 \frac{\text{lbm}}{\text{ft}^3}\right) + (1.3 \text{ lbm})\left(5 \frac{\text{lbm}}{\text{ft}^3}\right)}{14 \text{ lbm}} = 8.9 \text{ lbm/ft}^3$$

The volume at 40% moisture is

$$V_{40\%,moist} = \frac{m_{total}}{\rho_{ave} 0.4}$$
$$= \frac{14 \text{ lbm}}{\left(8.9 \frac{\text{lbm}}{\text{ft}^3}\right)(0.4)}$$
$$= 3.9 \text{ ft}^3/100 \text{ lbm waste}$$

The potential revenue is

$$\text{revenue} = (100{,}000 \text{ people})\left(5 \frac{\text{lbm waste}}{\text{person-day}}\right)$$
$$\times \left(\frac{3.9 \text{ ft}^3 \text{ mulch}}{100 \text{ lbm waste}}\right)\left(\frac{1 \text{ yd}^3}{27 \text{ ft}^3}\right)$$
$$\times \left(365 \frac{\text{day}}{\text{yr}}\right)\left(\frac{\$10}{\text{yd}^3}\right)$$
$$= \$2.6 \times 10^6/\text{yr} \quad (\$2.6 \text{ million/yr})$$

(C) This incorrect solution uses dry volume instead of wet volume at 40% moisture and then tries to convert to wet volume by dividing by the moisture fraction. Other assumptions are unchanged from the correct solution.

component	discarded mass, $m_{discarded}$ (lbm)	discarded moisture (%)	dry mass, m_{dry} (lbm)	dry density, $\rho_{dry,discarded}$ (lbm/ft³)	volume at 40% moisture, $V_{40\%,moist}$ (ft³)
garden	25	60	10	7	1.4
food	9	70	2.7	18	0.15
paper	38	6	36	5	7.2

The dry volume is

$$V_{dry} = \frac{m_{dry}}{\rho_{dry,discarded}}$$

The paper volume is

$$V_{paper} = (1.4 \text{ ft}^3 + 0.15 \text{ ft}^3)(0.1) = 0.16 \text{ ft}^3 < 7.2 \text{ ft}^3$$

Paper is not limiting.

The dry mulch volume per 100 lbm sample is

$$V_{\text{mulch,dry}} = 1.4 \text{ ft}^3 + 0.15 \text{ ft}^3 + 0.16 \text{ ft}^3 = 1.7 \text{ ft}^3$$

The volume at 40% moisture is

$$V_{40\%,\text{moist}} = \frac{1.7 \text{ ft}^3}{0.4} = 4.3 \text{ ft}^3$$

The potential revenue is

$$\begin{aligned}\text{revenue} &= (100{,}000 \text{ people})\left(5 \ \frac{\text{lbm waste}}{\text{person-day}}\right)\\ &\quad \times \left(\frac{4.3 \text{ ft}^3 \text{ mulch}}{100 \text{ lbm waste}}\right)\left(\frac{1 \text{ yd}^3}{27 \text{ yd}^3}\right)\\ &\quad \times \left(365 \ \frac{\text{day}}{\text{yr}}\right)\left(\frac{\$10}{\text{yd}^3}\right)\\ &= \$2.9 \times 10^6/\text{yr} \quad (\$2.9 \text{ million/yr})\end{aligned}$$

(D) This incorrect solution calculates dry mass by multiplying by the discarded moisture fraction instead of by the discarded dry mass fraction. Other assumptions are unchanged from the correct solution.

component	discarded mass, $m_{\text{discarded}}$ (lbm)	discarded moisture (%)	dry discarded mass, m_{dry} (lbm)	dry density, $\rho_{\text{dry,discarded}}$ (lbm/ft³)	volume at 40% moisture, $V_{40\%,\text{moist}}$ (ft³)
garden	25	60	15	7	3.6
food	9	70	6.3	18	0.58
paper	38	6	2.3	5	0.77

The dry mass is

$$m_{\text{dry}} = \frac{(m_{\text{discarded}})(\text{discarded moisture in \%})}{100\%}$$

The volume at 40% moisture is

$$V_{40\%,\text{moist}} = \frac{m_{\text{dry}}}{\rho_{\text{dry,discarded}}(1-0.4)}$$

The paper volume is

$$V_{\text{paper}} = (3.6 \text{ ft}^3 + 0.58 \text{ ft}^3)(0.1) = 0.42 \text{ ft}^3 < 0.77 \text{ ft}^3$$

Paper is not limiting.

The mulch volume per 100 lbm waste is

$$V_{\text{mulch}} = 3.6 \text{ ft}^3 + 0.58 \text{ ft}^3 + 0.42 \text{ ft}^3 = 4.6 \text{ ft}^3$$

The potential revenue is

$$\begin{aligned}\text{revenue} &= (100{,}000 \text{ people})\left(5 \ \frac{\text{lbm waste}}{\text{person-day}}\right)\\ &\quad \times \left(\frac{4.6 \text{ ft}^3 \text{ mulch}}{100 \text{ lbm waste}}\right)\left(\frac{1 \text{ yd}^3}{27 \text{ yd}^3}\right)\\ &\quad \times \left(365 \ \frac{\text{day}}{\text{yr}}\right)\left(\frac{\$10}{\text{yd}^3}\right)\\ &= \$3.1 \times 10^6/\text{yr} \quad (\$3.1 \text{ million/yr})\end{aligned}$$

SOLUTION 64

component	average of total discarded (%)	average mass discarded, m_{ave} (kg)	average curbside density, ρ_{ave} (kg/m³)	average volume discarded, V_{ave} (m³)
newspaper	16	524	200	2.62
aluminum	1.3	42.6	240	0.18
steel	1.9	62.2	405	0.15
plastic	5.7	187	265	0.71
glass	4.1	134	180	0.74
				4.4

The average mass discarded is

$$\begin{aligned}m_{\text{ave}} &= \left(\frac{\text{discarded\%}}{100\%}\right)\left(1.3 \ \frac{\text{kg}}{\text{person} \cdot \text{d}}\right)(7 \text{ d})\\ &\quad \times \left(3.6 \ \frac{\text{persons}}{\text{stop}}\right)(100 \text{ stops})\end{aligned}$$

The average volume discarded is

$$V_{\text{ave}} = \frac{m_{\text{ave}}}{\rho_{\text{ave}}}$$

From the table, the total volume discarded is 4.4 m³/wk.

The truck volume required is 5 m³.

The answer is (D).

Why Other Options Are Wrong

(A) This solution is incorrect because it calculates the average mass for one day instead of for seven days.

component	average of total discarded (%)	average mass discarded, m_{ave} (kg)	average curbside density, ρ_{ave} (kg/m³)	average volume discarded, V_{ave} (m³)
newspaper	16	75	200	0.38
aluminum	1.3	6.1	240	0.025
steel	1.9	8.9	405	0.022
plastic	5.7	27	265	0.11
glass	4.1	19	180	0.11
				0.65

The average mass discarded is

$$\begin{aligned}m_{\text{ave}} &= \left(\frac{\text{discarded \%}}{100\%}\right)(1.3 \text{ kg/person})\\ &\quad \times (3.6 \text{ persons/stop})(100 \text{ stops})\end{aligned}$$

The average volume discarded is

$$V_{ave} = \frac{m_{ave}}{\rho_{ave}}$$

From the table, the total volume discarded is 0.65 m^3.

The truck volume required is 1 m^3.

(B) This solution is incorrect because it fails to include the number of persons per stop in the average mass discarded calculation.

component	average of total discarded (%)	average mass discarded, m_{ave} (kg)	average curbside density, ρ_{ave} (kg/m³)	average volume discarded, V_{ave} (m³)
newspaper	16	146	200	0.73
aluminum	1.3	12.0	240	0.050
steel	1.9	17.3	405	0.043
plastic	5.7	51.9	265	0.20
glass	4.1	37.3	180	0.21
				1.23

The average mass discarded is

$$m_{ave} = \left(\frac{\text{discarded \%}}{100\%}\right)(1.3 \text{ kg/d})(7 \text{ d/stop})$$
$$\times (100 \text{ stops})$$

The average volume discarded is

$$V_{ave} = \frac{m_{ave}}{\rho_{ave}}$$

From the table, the total volume discarded is 1.23 m^3.

The truck volume required is 2 m^3.

(C) This solution is incorrect because it bases the volume on the average curbside density.

component	average of total discarded (%)	average mass discarded, m_{ave} (kg)	average curbside density, ρ_{ave} (kg/m³)
newspaper	16	524	200
aluminum	1.3	42.6	240
steel	1.9	62.2	405
plastic	5.7	187	265
glass	4.1	134	180
		950	1290

The average mass discarded is

$$m_{ave} = \left(\frac{\text{discarded \%}}{100\%}\right)\left(1.3 \frac{\text{kg}}{\text{person}\cdot\text{d}}\right)(7 \text{ d})$$
$$\times \left(3.6 \frac{\text{persons}}{\text{stop}}\right)(100 \text{ stops})$$

The average density is

$$\rho_{ave} = \frac{1290 \frac{\text{kg}}{\text{m}^3}}{5} = 258 \text{ kg/m}^3$$

The total volume discarded is

$$V_{total} = \frac{950 \text{ kg}}{258 \frac{\text{kg}}{\text{m}^3}} = 3.7 \text{ m}^3/\text{week}$$

The truck volume required is 4 m^3.

SOLUTION 65

The number of dumpsters per load is

$$\frac{\left(8 \frac{\text{yd}^3}{\text{load}}\right)\left(1200 \frac{\text{lbm}}{\text{yd}^3}\right)}{\left(3 \frac{\text{yd}^3}{\text{dumpster}}\right)\left(\frac{1000 \text{ lbm}}{3 \text{ yd}^3}\right)}$$
$$= 9.6 \text{ dumpsters/load} \quad (9 \text{ dumpsters/load})$$

The time required per load is

$$\left(2 \frac{\text{min}}{\text{dumpster}} + 6 \frac{\text{min}}{\text{dumpster}}\right)\left(9 \frac{\text{dumpsters}}{\text{load}}\right)$$
$$+ 38 \frac{\text{min}}{\text{load}}$$
$$= 110 \text{ min/load}$$

The time available for collection in a single day is

$$\left(8 \frac{\text{hr}}{\text{day}}\right)\left(60 \frac{\text{min}}{\text{hr}}\right) = 480 \text{ min/day}$$

The number of loads per day is

$$\frac{480 \frac{\text{min}}{\text{day}}}{110 \frac{\text{min}}{\text{load}}} = 4.4 \text{ loads/day} \quad (4 \text{ loads/day})$$

The number of dumpsters emptied in a single day is

$$\left(9 \frac{\text{dumpsters}}{\text{load}}\right)\left(4 \frac{\text{loads}}{\text{day}}\right) = 36 \text{ dumpsters/day}$$

The answer is (C).

Why Other Options Are Wrong

(A) This incorrect choice does not include compaction in calculating the number of dumpsters per truckload.

The number of dumpsters per load is

$$\left(8 \, \frac{\text{yd}^3}{\text{load}}\right)\left(\frac{1 \text{ dumpster}}{3 \text{ yd}^3}\right)$$
$$= 2.7 \text{ dumpsters/load} \quad (2 \text{ dumpsters/load})$$

The time required per load is

$$\left(\frac{(2 \text{ min} + 6 \text{ min})}{\text{dumpster}}\right)\left(2 \, \frac{\text{dumpsters}}{\text{load}}\right) + 38 \, \frac{\text{min}}{\text{load}}$$
$$= 54 \text{ min/load}$$

The time available for collection in a single day is

$$\left(8 \, \frac{\text{hr}}{\text{day}}\right)\left(60 \, \frac{\text{min}}{\text{hr}}\right) = 480 \text{ min/day}$$

The number of loads per day is

$$\frac{480 \, \frac{\text{min}}{\text{day}}}{54 \, \frac{\text{min}}{\text{load}}} = 8.9 \text{ loads/day} \quad (9 \text{ loads/day})$$

The number of dumpsters emptied in a single day is

$$\left(2 \, \frac{\text{dumpsters}}{\text{load}}\right)\left(9 \, \frac{\text{loads}}{\text{day}}\right) = 18 \text{ dumpsters/day}$$

(B) This incorrect choice calculates as discarded density for 1 yd³ instead of the dumpster 3 yd³ capacity. The number of dumpsters per load is

$$\frac{\left(8 \, \frac{\text{yd}^3}{\text{load}}\right)\left(1200 \, \frac{\text{lbm}}{\text{yd}^3}\right)}{\left(3 \, \frac{\text{yd}^3}{\text{dumpster}}\right)\left(1000 \, \frac{\text{lbm}}{\text{yd}^3}\right)}$$
$$= 3.2 \text{ dumpsters/load} \quad (3 \text{ dumpsters/load})$$

The time required per load is

$$\left(\frac{(2 \text{ min} + 6 \text{ min})}{\text{dumpster}}\right)\left(3 \, \frac{\text{dumpsters}}{\text{load}}\right) + 38 \, \frac{\text{min}}{\text{load}}$$
$$= 62 \text{ min/load}$$

The time available for collection in a single day is

$$\left(8 \, \frac{\text{hr}}{\text{day}}\right)\left(60 \, \frac{\text{min}}{\text{hr}}\right) = 480 \text{ min/day}$$

The number of loads per day is

$$\frac{480 \, \frac{\text{min}}{\text{day}}}{62 \, \frac{\text{min}}{\text{load}}} = 7.7 \text{ loads/day} \quad (7 \text{ loads/day})$$

The number of dumpsters emptied in a single day is

$$\left(3 \, \frac{\text{dumpsters}}{\text{load}}\right)\left(7 \, \frac{\text{loads}}{\text{d}}\right) = 21 \text{ dumpsters/day}$$

(D) This incorrect choice does not consider all of the relevant given information.

The time available for collection in a single day is

$$\left(8 \, \frac{\text{hr}}{\text{day}}\right)\left(60 \, \frac{\text{min}}{\text{hr}}\right) = 480 \text{ min/day}$$

The time dedicated to collection, excluding travel, is

$$480 \, \frac{\text{min}}{\text{day}} - 38 \, \frac{\text{min in a.m.}}{\text{day}} - 38 \, \frac{\text{min in p.m.}}{\text{day}}$$
$$= 404 \text{ min time/day}$$

The number of dumpsters emptied in a single day is

$$\frac{404 \, \frac{\text{min collection time}}{\text{day}}}{8 \, \frac{\text{min}}{\text{dumpster}}}$$
$$= 50.5 \text{ dumpsters/day} \quad (50 \text{ dumpsters/day})$$

SOLUTION 66

$\dot{m}$	waste mass flow rate	kg/d
q	net heat	kJ/d
q_a	heat loss to ash	kJ/d
q_r	heat loss to radiation	kJ/d
q_T	total heat	kJ/d
q_v	latent heat of vaporization of water	2420 kJ/kg
q_w	heat loss to water	kJ/d

$$\dot{m} = \left(72 \, \frac{\text{tons}}{\text{day}}\right)\left(2000 \, \frac{\text{lbm}}{\text{ton}}\right)\left(\frac{1 \text{ kg}}{2.204 \text{ lbm}}\right)$$
$$= 65\,336 \text{ kg/d}$$

$$q_T = \left(65\,336 \, \frac{\text{kg}}{\text{d}}\right)\left(11\,000 \, \frac{\text{kJ}}{\text{kg}}\right)$$
$$= 7.2 \times 10^8 \text{ kJ/d}$$

$$q_a = \left(65\,336 \, \frac{\text{kg}}{\text{d}}\right)(0.12)\left(400 \, \frac{\text{kJ}}{\text{kg}}\right)$$
$$= 3.1 \times 10^6 \text{ kJ/d}$$

$$q_w = \left(65\,336 \, \frac{\text{kg}}{\text{d}}\right)(0.48)\left(2420 \, \frac{\text{kJ}}{\text{kg}}\right)$$
$$= 7.6 \times 10^7 \text{ kJ/d}$$

$$q_r = \left(65\,336 \, \frac{\text{kg}}{\text{d}}\right)\left(0.0035 \, \frac{\text{kJ}}{\text{kg}}\right)$$
$$= 229 \text{ kJ/d} \quad (\text{negligible})$$

$$q = q_T - q_a - q_w - q_r$$
$$= 7.2 \times 10^8 \, \frac{\text{kJ}}{\text{d}} - 3.1 \times 10^6 \, \frac{\text{kJ}}{\text{d}} - 7.6 \times 10^7 \, \frac{\text{kJ}}{\text{d}}$$
$$= 6.4 \times 10^8 \, \text{kJ/d}$$

The answer is (B).

Why Other Options Are Wrong

(A) This incorrect solution calculates net heat by subtracting the ash and total water fractions from the total waste mass. Other definitions and equations are unchanged from the correct solution.

$$\dot{m} = \left(72 \, \frac{\text{tons}}{\text{day}}\right)\left(2000 \, \frac{\text{lbm}}{\text{ton}}\right)\left(\frac{1 \, \text{kg}}{2.204 \, \text{lbm}}\right)$$
$$\times (1 - 0.12 - 0.48)$$
$$= 26\,134 \, \text{kg/d}$$
$$q = \left(26\,134 \, \frac{\text{kg}}{\text{d}}\right)\left(11\,000 \, \frac{\text{kJ}}{\text{kg}}\right)$$
$$= 2.9 \times 10^8 \, \text{kJ/d}$$

(C) This incorrect solution uses the specific heat instead of the heat of vaporization of water to calculate the heat loss to water. Other definitions and equations are unchanged from the correct solution.

$$\dot{m} = \left(72 \, \frac{\text{tons}}{\text{day}}\right)\left(2000 \, \frac{\text{lbm}}{\text{ton}}\right)\left(\frac{1 \, \text{kg}}{2.204 \, \text{lbm}}\right)$$
$$= 65\,336 \, \text{kg/d}$$
$$q_T = \left(65\,336 \, \frac{\text{kg}}{\text{d}}\right)\left(11\,000 \, \frac{\text{kJ}}{\text{kg}}\right)$$
$$= 7.2 \times 10^8 \, \text{kJ/d}$$
$$q_a = \left(65\,336 \, \frac{\text{kg}}{\text{d}}\right)(0.12)\left(400 \, \frac{\text{kJ}}{\text{kg}}\right)$$
$$= 3.1 \times 10^6 \, \text{kJ/d}$$

Use 4.187 kJ/kg°C for the specific heat of water and assume that the water temperature in the waste is initially at 25°C.

$$q_w = \left(65\,336 \, \frac{\text{kg}}{\text{d}}\right)(0.48)\left(4.187 \, \frac{\text{kJ}}{\text{kg°C}}\right)$$
$$\times (100°\text{C} - 25°\text{C})$$
$$= 9.8 \times 10^6 \, \text{kJ/d}$$
$$q_r = \left(65\,336 \, \frac{\text{kg}}{\text{d}}\right)\left(0.0035 \, \frac{\text{kJ}}{\text{kg}}\right)$$
$$= 229 \, \text{kJ/d} \quad (\text{negligible})$$
$$q = 7.2 \times 10^8 \, \frac{\text{kJ}}{\text{d}} - 3.1 \times 10^6 \, \frac{\text{kJ}}{\text{d}} - 9.8 \times 10^6 \, \frac{\text{kJ}}{\text{d}}$$
$$= 7.1 \times 10^8 \, \text{kJ/d}$$

(D) This incorrect solution uses an improper factor for converting tons to kilograms. Other definitions and equations are unchanged from the correct solution.

$$\dot{m} = \left(72 \, \frac{\text{tons}}{\text{day}}\right)\left(2000 \, \frac{\text{kg}}{\text{ton}}\right)$$
$$= 144\,000 \, \text{kg/d}$$
$$q_T = \left(144\,000 \, \frac{\text{kg}}{\text{d}}\right)\left(11\,000 \, \frac{\text{kJ}}{\text{kg}}\right)$$
$$= 1.6 \times 10^9 \, \text{kJ/d}$$
$$q_a = \left(144\,000 \, \frac{\text{kg}}{\text{d}}\right)(0.12)\left(400 \, \frac{\text{kJ}}{\text{kg}}\right)$$
$$= 7.0 \times 10^6 \, \text{kJ/d}$$
$$q_w = \left(144\,000 \, \frac{\text{kg}}{\text{d}}\right)(0.48)\left(2420 \, \frac{\text{kJ}}{\text{kg}}\right)$$
$$= 1.7 \times 10^8 \, \text{kJ/d}$$
$$q_r = \left(144\,000 \, \frac{\text{kg}}{\text{d}}\right)\left(0.0035 \, \frac{\text{kJ}}{\text{kg}}\right)$$
$$= 504 \, \text{kJ/d} \quad (\text{negligible})$$
$$q = 1.6 \times 10^9 \, \frac{\text{kJ}}{\text{d}} - 7.0 \times 10^6 \, \frac{\text{kJ}}{\text{d}} - 1.7 \times 10^8 \, \frac{\text{kJ}}{\text{d}}$$
$$= 1.4 \times 10^9 \, \text{kJ/d}$$

SOLUTION 67

The waste mass per week is

$$(230{,}000 \, \text{people})\left(2.5 \, \frac{\text{lbm}}{\text{person-day}}\right)\left(7 \, \frac{\text{days}}{\text{wk}}\right)$$
$$= 4.025 \times 10^6 \, \text{lbm/wk}$$

The machine capacity per week is

$$\left(3500 \, \frac{\text{lbm}}{\text{bale}}\right)\left(16 \, \frac{\text{bales}}{\text{hr}}\right)\left(40 \, \frac{\text{hr}}{\text{machine-wk}}\right)$$
$$= 2.24 \times 10^6 \, \text{lbm/machine-wk}$$

The minimum number of machines required is

$$\frac{4.025 \times 10^6 \, \frac{\text{lbm}}{\text{wk}}}{2.24 \times 10^6 \, \frac{\text{lbm}}{\text{machine-wk}}}$$
$$= 1.8 \, \text{machines} \quad (2 \, \text{machines})$$

The answer is (B).

Why Other Options Are Wrong

(A) This incorrect solution calculates the waste generation rate for a single day instead of a seven day week and then uses a 24 hr day to calculate machine capacity.

The waste mass per day is

$$(230{,}000 \text{ people})\left(2.5 \ \frac{\text{lbm}}{\text{person-day}}\right)$$
$$= 575{,}000 \ \text{lbm/day}$$

The machine capacity per day is

$$\left(3500 \ \frac{\text{lbm}}{\text{bale}}\right)\left(16 \ \frac{\text{bales}}{\text{machine-hr}}\right)\left(24 \ \frac{\text{hr}}{\text{day}}\right)$$
$$= 1{,}344{,}000 \ \text{lbm/machine-day}$$

The minimum number of machines required is

$$\frac{575{,}000 \ \frac{\text{lbm}}{\text{day}}}{1{,}344{,}000 \ \frac{\text{lbm}}{\text{machine-day}}}$$
$$= 0.43 \ \text{machine} \quad (1 \ \text{machine})$$

(C) This incorrect solution calculates the waste generation rate for a five day instead of seven day week and divides the 40 hr operating time by five days.

The waste mass per day is

$$(230{,}000 \text{ people})\left(2.5 \ \frac{\text{lbm}}{\text{person-day}}\right)(5 \text{ days})$$
$$= 2.88 \times 10^6 \ \text{lbm/day}$$

The machine capacity per day is

$$\left(3500 \ \frac{\text{lbm}}{\text{bale}}\right)\left(16 \ \frac{\text{bales}}{\text{machine-hr}}\right)\left(\frac{40 \ \text{hr}}{5 \ \text{days}}\right)$$
$$= 448{,}000 \ \text{lbm/machine-day}$$

The minimum number of machines required is

$$\frac{2.88 \times 10^6 \ \frac{\text{lbm}}{\text{day}}}{448{,}000 \ \frac{\text{lbm}}{\text{machine-day}}} = 6.4 \ \text{machines} \quad (7 \ \text{machines})$$

(D) This incorrect solution divides the 40 hr week by five days when calculating baling machine capacity.

The waste mass per day is

$$(230{,}000 \text{ people})\left(2.5 \ \frac{\text{lbm}}{\text{person-day}}\right)(7 \text{ days})$$
$$= 4.025 \times 10^6 \ \text{lbm/day}$$

The machine capacity per day is

$$\left(3500 \ \frac{\text{lbm}}{\text{bale}}\right)\left(16 \ \frac{\text{bales}}{\text{machine-hr}}\right)\left(\frac{40 \ \text{hr}}{5 \ \text{days}}\right)$$
$$= 448{,}000 \ \text{lbm/machine-day}$$

The minimum of machines required is

$$\frac{4.025 \times 10^6 \ \frac{\text{lbm}}{\text{day}}}{448{,}000 \ \frac{\text{lbm}}{\text{machine-day}}} = 9.0 \ \text{machines} \quad (9 \ \text{machines})$$

SOLUTION 68

C_f adsorber effluent concentration   mg/L

$$C_f = (1 - 0.99)\left(148 \ \frac{\text{mg}}{\text{L}}\right) = 1.48 \ \text{mg/L}$$

C_o initial concentration   mg/L
k isotherm intercept   mg/g
$1/n$ isotherm slope   0.45
M granular activated carbon (GAC) mass-use rate   g/L

$$M = \frac{C_o - C_f}{kC_f^{1/n}} = \frac{148 \ \frac{\text{mg}}{\text{L}} - 1.48 \ \frac{\text{mg}}{\text{L}}}{\left(220 \ \frac{\text{mg}}{\text{g}}\right)\left(1.48 \ \frac{\text{mg}}{\text{L}}\right)^{0.45}}$$
$$= 0.56 \ \text{g GAC/L water treated}$$

$\dot{m}$ mass flow rate   lbm/day
$\dot{V}$ volumetric flow rate   gal/day

The daily GAC use rate is

$$\dot{m} = M\dot{V}$$
$$= \left(0.56 \ \frac{\text{g}}{\text{L}}\right)\left(2.5 \times 10^5 \ \frac{\text{gal}}{\text{day}}\right)\left(3.785 \ \frac{\text{L}}{\text{gal}}\right)$$
$$\times \left(\frac{2.204 \ \text{lbm}}{1000 \ \text{g}}\right)$$
$$= 1168 \ \text{lbm/day}$$

Standard sizes of adsorption vessels are 2000 lbm, 4000 lbm, 10,000 lbm, and 20,000 lbm.

The number of days to saturation is

$$\frac{20{,}000 \ \text{lbm}}{1168 \ \frac{\text{lbm}}{\text{day}}} = 17 \ \text{day} > 14 \ \text{day}$$

Only the 20,000 lbm vessel will provide a minimum GAC change-out period of 14 days.

The answer is (D).

Why Other Options Are Wrong

(A) This incorrect solution miscalculates the effluent concentration. Other definitions, equations, and standard vessel sizes are the same as for the correct solution.

$$C_f = (0.99)\left(148 \ \frac{\text{mg}}{\text{L}}\right) = 146.52 \ \text{mg/L}$$

$$M = \frac{148\ \frac{\text{mg}}{\text{L}} - 146.52\ \frac{\text{mg}}{\text{L}}}{\left(220\ \frac{\text{mg}}{\text{g}}\right)\left(146.52\ \frac{\text{mg}}{\text{L}}\right)^{0.45}}$$
$$= 0.00071\ \text{g GAC/L water treated}$$

The daily GAC use rate is

$$\left(0.00071\ \frac{\text{g GAC}}{\text{L}}\right)\left(2.5 \times 10^5\ \frac{\text{gal}}{\text{day}}\right)$$
$$\times \left(3.785\ \frac{\text{L}}{\text{gal}}\right)\left(\frac{2.204\ \text{lbm}}{1000\ \text{g}}\right)$$
$$= 1.5\ \text{lbm/day}$$

The number of days to saturation is

$$\frac{2000\ \text{lbm}}{1.5\ \frac{\text{lbm}}{\text{day}}} = 1333\ \text{day} > 14\ \text{day}$$

The 2000 lbm vessel is the smallest vessel that will provide a minimum GAC change-out period of 14 days.

(B) This incorrect solution confuses the units in the pounds mass to grams conversion factor. Other definitions, equations, and standard vessel sizes are the same as the correct solution.

$$C_f = (1 - 0.99)\left(148\ \frac{\text{mg}}{\text{L}}\right)$$
$$= 1.48\ \text{mg/L}$$
$$M = \frac{148\ \frac{\text{mg}}{\text{L}} - 1.48\ \frac{\text{mg}}{\text{L}}}{\left(220\ \frac{\text{mg}}{\text{g}}\right)\left(1.48\ \frac{\text{mg}}{\text{L}}\right)^{0.45}}$$
$$= 0.56\ \text{g GAC/L water treated}$$

The daily GAC use rate is

$$\left(0.56\ \frac{\text{g GAC}}{\text{L}}\right)\left(2.5 \times 10^5\ \frac{\text{gal}}{\text{day}}\right)$$
$$\times \left(3.785\ \frac{\text{L}}{\text{gal}}\right)\left(\frac{1\ \text{lbm}}{2204\ \text{g}}\right)$$
$$= 240\ \text{lbm/day}$$

The number of days to saturation is

$$\frac{4000\ \text{lbm}}{240\ \frac{\text{lbm}}{\text{day}}} = 17\ \text{day} > 14\ \text{day}$$

The 4000 lbm vessel is the smallest vessel that will provide a minimum GAC change-out period of 14 days.

(C) This incorrect solution uses the inverse slope in the isotherm equation. Other definitions, equations, and standard vessel sizes are the same as the correct solution.

$$C_f = (1 - 0.99)\left(148\ \frac{\text{mg}}{\text{L}}\right)$$
$$= 1.48\ \text{mg/L}$$
$$M = \frac{148\ \frac{\text{mg}}{\text{L}} - 1.48\ \frac{\text{mg}}{\text{L}}}{\left(220\ \frac{\text{mg}}{\text{g}}\right)\left(1.48\ \frac{\text{mg}}{\text{L}}\right)^{1/0.45}}$$
$$= 0.28\ \text{g GAC/L water treated}$$

The daily GAC use rate is

$$\left(0.28\ \frac{\text{g GAC}}{\text{L}}\right)\left(2.5 \times 10^5\ \frac{\text{gal}}{\text{day}}\right)$$
$$\times \left(3.785\ \frac{\text{L}}{\text{gal}}\right)\left(\frac{2.204\ \text{lbm}}{1000\ \text{g}}\right)$$
$$= 584\ \text{lbm/day}$$

The number of days to saturation is

$$\frac{10{,}000\ \text{lbm}}{584\ \frac{\text{lbm}}{\text{day}}} = 17\ \text{day} > 14\ \text{day}$$

The 10,000 lbm vessel is the smallest vessel that will provide a minimum GAC change-out period of 14 days.

SOLUTION 69

The cations will be removed by the cation exchanger. In the cation exchanger, Zn, Cd, and Ni are exchanged for H^+. Size the exchanger for one day of operation.

The equivalents of cations removed per liter of water treated are

$$\frac{\left(31\ \frac{\text{mg Cd}}{\text{L}}\right)\left(2\ \frac{\text{meq}}{\text{mol}}\right)}{112\ \frac{\text{g Cd}}{\text{mol}}} + \frac{\left(13\ \frac{\text{mg Zn}}{\text{L}}\right)\left(2\ \frac{\text{meq}}{\text{mol}}\right)}{65\ \frac{\text{g Zn}}{\text{mol}}}$$
$$+ \frac{\left(21\ \frac{\text{mg Ni}}{\text{L}}\right)\left(2\ \frac{\text{meq}}{\text{mol}}\right)}{59\ \frac{\text{g Ni}}{\text{mol}}}$$
$$= 1.67\ \frac{\text{mg·meq}}{\text{g·L}}$$
$$= 0.00167\ \text{meq/L}$$

The resin volume is

$$\frac{\left(164\ \frac{\text{m}^3}{\text{d}}\right)\left(0.00167\ \frac{\text{meq}}{\text{L}}\right)}{1.5\ \frac{\text{equiv}}{\text{L resin}}} = 0.18\ \text{m}^3\ \text{resin/d}$$

The regeneration volume is

$$\frac{\left(192 \frac{\text{kg H}_2\text{SO}_4}{\text{m}^3}\right)\left(0.18 \frac{\text{m}^3}{\text{d}}\right)\left(1000 \frac{\text{g}}{\text{kg}}\right)}{\left(1.44 \frac{\text{g}}{\text{cm}^3}\right)\left(\frac{5 \text{ kg H}_2\text{SO}_4}{100 \text{ kg solution}}\right)\left(10^6 \frac{\text{cm}^3}{\text{m}^3}\right)}$$

$$= 0.48 \text{ m}^3 \text{ solution/d}$$

The regeneration period is

$$\frac{0.48 \frac{\text{m}^3 \text{ solution}}{\text{d}}}{\left(0.020 \frac{\text{m}^3 \text{ solution}}{\text{m}^3 \text{ resin} \cdot \text{min}}\right)\left(0.18 \frac{\text{m}^3 \text{ resin}}{\text{d}}\right)}$$

$$= 133 \text{ min} \quad (130 \text{ min})$$

The answer is (D).

Why Other Options Are Wrong

(A) This incorrect solution ignores the regeneration solution concentration. Other assumptions are unchanged from the correct solution.

The equivalents of cations removed per liter of water treated are

$$\frac{\left(31 \frac{\text{mg Cd}}{\text{L}}\right)\left(2 \frac{\text{meq}}{\text{mol}}\right)}{112 \frac{\text{g Cd}}{\text{mol}}} + \frac{\left(13 \frac{\text{mg Zn}}{\text{L}}\right)\left(2 \frac{\text{meq}}{\text{mol}}\right)}{65 \frac{\text{g Zn}}{\text{mol}}}$$

$$+ \frac{\left(21 \frac{\text{mg Ni}}{\text{L}}\right)\left(2 \frac{\text{meq}}{\text{mol}}\right)}{59 \frac{\text{g Ni}}{\text{mol}}}$$

$$= 1.67 \frac{\text{mg} \cdot \text{meq}}{\text{g} \cdot \text{L}}$$

$$= 0.00167 \text{ meq/L}$$

The resin volume is

$$\frac{\left(164 \frac{\text{m}^3}{\text{d}}\right)\left(0.00167 \frac{\text{meq}}{\text{L}}\right)}{1.5 \frac{\text{meq}}{\text{L}}} = 0.18 \text{ m}^3 \text{ resin/d}$$

The regeneration volume is

$$\frac{\left(192 \frac{\text{kg H}_2\text{SO}_4}{\text{m}^3}\right)\left(0.18 \frac{\text{m}^3}{\text{d}}\right)\left(1000 \frac{\text{g}}{\text{kg}}\right)}{\left(1.44 \frac{\text{g}}{\text{cm}^3}\right)\left(10^6 \frac{\text{cm}^3}{\text{m}^3}\right)}$$

$$= 0.024 \text{ m}^3 \text{ solution/d}$$

The regeneration period is

$$\frac{0.024 \frac{\text{m}^3 \text{ solution}}{\text{d}}}{\left(0.020 \frac{\text{m}^3 \text{ solution}}{\text{m}^3 \text{ resin} \cdot \text{min}}\right)\left(0.18 \frac{\text{m}^3 \text{ resin}}{\text{d}}\right)} = 7 \text{ min}$$

(B) This incorrect solution confuses the parameters in the regeneration period calculation. Other assumptions are unchanged from the correct solution.

The equivalents of cations removed per liter of water treated are

$$\frac{\left(31 \frac{\text{mg Cd}}{\text{L}}\right)\left(2 \frac{\text{meq}}{\text{mol}}\right)}{112 \frac{\text{g Cd}}{\text{mol}}} + \frac{\left(13 \frac{\text{mg Zn}}{\text{L}}\right)\left(2 \frac{\text{meq}}{\text{mol}}\right)}{65 \frac{\text{g Zn}}{\text{mol}}}$$

$$+ \frac{\left(21 \frac{\text{mg Ni}}{\text{L}}\right)\left(2 \frac{\text{meq}}{\text{mol}}\right)}{59 \frac{\text{g Ni}}{\text{mol}}}$$

$$= 1.67 \frac{\text{mg} \cdot \text{meq}}{\text{g} \cdot \text{L}}$$

$$= 0.00167 \text{ meq/L}$$

The resin volume is

$$\frac{\left(164 \frac{\text{m}^3}{\text{d}}\right)\left(0.00167 \frac{\text{meq}}{\text{L}}\right)}{1.5 \frac{\text{meq}}{\text{L}}} = 0.18 \text{ m}^3 \text{ resin/d}$$

The regeneration volume is

$$\frac{\left(192 \frac{\text{kg H}_2\text{SO}_4}{\text{m}^3}\right)\left(0.18 \frac{\text{m}^3}{\text{d}}\right)\left(1000 \frac{\text{g}}{\text{kg}}\right)}{\left(1.44 \frac{\text{g}}{\text{cm}^3}\right)\left(5 \frac{\text{kg H}_2\text{SO}_4}{100 \text{ kg solution}}\right)\left(10^6 \frac{\text{cm}^3}{\text{m}^3}\right)}$$

$$= 0.48 \text{ m}^3 \text{ solution/d}$$

The regeneration period is

$$\frac{\left(0.18 \frac{\text{m}^3}{\text{d}}\right)\left(0.020 \frac{\text{m}^3}{\text{m}^3 \cdot \text{min}}\right)\left(1440 \frac{\text{min}}{\text{d}}\right)}{0.48 \frac{\text{m}^3}{\text{d}}} = 11 \text{ min}$$

Note that units do not work.

(C) This incorrect solution uses the anion exchanger instead of the cation exchanger characteristics. Other assumptions are unchanged from the correct solution.

Assume that the cations will be removed by the anion exchanger.

The equivalents of cations removed per liter of water treated are

$$\frac{\left(31 \frac{\text{mg Cd}}{\text{L}}\right)\left(2 \frac{\text{meq}}{\text{mol}}\right)}{112 \frac{\text{g Cd}}{\text{mol}}} + \frac{\left(13 \frac{\text{mg Zn}}{\text{L}}\right)\left(2 \frac{\text{meq}}{\text{mol}}\right)}{65 \frac{\text{g Zn}}{\text{mol}}}$$

$$+ \frac{\left(21 \frac{\text{mg Ni}}{\text{L}}\right)\left(2 \frac{\text{meq}}{\text{mol}}\right)}{59 \frac{\text{g Ni}}{\text{mol}}}$$

$$= 1.67 \frac{\text{mg·meq}}{\text{g·L}}$$

$$= 0.00167 \text{ meq/L}$$

The resin volume is

$$\frac{\left(164 \frac{\text{m}^3}{\text{d}}\right)\left(0.00167 \frac{\text{meq}}{\text{L}}\right)}{3.7 \frac{\text{meq}}{\text{L}}} = 0.074 \text{ m}^3 \text{ resin/d}$$

The regeneration volume is

$$\frac{\left(76 \frac{\text{kg NaOH}}{\text{m}^3}\right)\left(0.074 \frac{\text{m}^3}{\text{d}}\right)\left(1000 \frac{\text{g}}{\text{kg}}\right)}{\left(1.15 \frac{\text{g}}{\text{cm}^3}\right)\left(10 \frac{\text{kg NaOH}}{100 \text{ kg solution}}\right)\left(10^6 \frac{\text{cm}^3}{\text{m}^3}\right)}$$

$$= 0.049 \text{ m}^3 \text{ solution/d}$$

The regeneration period is

$$\frac{0.049 \frac{\text{m}^3 \text{ solution}}{\text{d}}}{\left(0.020 \frac{\text{m}^3 \text{ solution}}{\text{m}^3 \text{ resin·min}}\right)\left(0.074 \frac{\text{m}^3 \text{ resin}}{\text{d}}\right)} = 33 \text{ min}$$

SOLUTION 70

Assume that the neutral solution has a pH of 7.0.

$$\text{pH} + \text{pOH} = 14$$

At pH of 1.6,

$$\text{pOH} = 14 - 1.6 = 12.4$$
$$[\text{OH}^-] = 10^{-12.4} \text{ mol/L} \quad (4.0 \times 10^{-13} \text{ mol/L})$$

At pH of 7.0,

$$\text{pOH} = 14 - 7.0 = 7.0$$
$$[\text{OH}^-] = 10^{-7.0} \text{ mol/L} \quad (1.0 \times 10^{-7} \text{ mol/L})$$

$$\text{OH}^- \text{ required} = 1.0 \times 10^{-7} \frac{\text{mol}}{\text{L}} - 4.0 \times 10^{-13}$$

$$= 1.0 \times 10^{-7} \text{ mol/L}$$

$$\text{NaOH} \rightarrow \text{Na}^+ + \text{OH}^-$$

One mole of NaOH dissociates to produce one mole of OH^-.

$$\text{NaOH required} = 1.0 \times 10^{-7} \text{ mol/L}$$

The daily sodium hydroxide feed rate is

$$\frac{\left(50{,}000 \frac{\text{gal waste}}{\text{day}}\right)\left(1.0 \times 10^{-7} \frac{\text{mol NaOH}}{\text{L waste}}\right)}{0.005 \frac{\text{meq}}{\text{L acid}}}$$
$$\times \left(1 \frac{\text{meq}}{\text{mol NaOH}}\right)\left(3785 \frac{\text{mL}}{\text{gal}}\right)$$

$$= 3.8 \times 10^3 \text{ mL/d}$$

The answer is (B).

Why Other Options Are Wrong

(A) This incorrect solution bases the hydroxide concentration on the pOH for a corresponding pH of 1.6 and does not consider the pOH under neutralized conditions. Other assumptions are unchanged from the correct solution.

$$\text{pH} + \text{pOH} = 14$$

At pH of 1.6,

$$\text{pOH} = 14 - 1.6 = 12.4$$
$$\text{OH}^- \text{ required} = 10^{-12.4} \text{ mol/L} = 4.0 \times 10^{-13} \text{ mol/L}$$
$$\text{NaOH} \rightarrow \text{Na}^+ + \text{OH}^-$$

One mole of NaOH dissociates to produce one mole of OH^-.

$$\text{NaOH required} = 4.0 \times 10^{-13} \text{ mol/L}$$

The daily sodium hydroxide feed rate is

$$\frac{\left(50{,}000 \frac{\text{gal waste}}{\text{day}}\right)\left(4.0 \times 10^{-13} \frac{\text{mol NaOH}}{\text{L waste}}\right)}{0.005 \frac{\text{equiv}}{\text{L acid}}}$$
$$\times \left(1 \frac{\text{equiv}}{\text{mol NaOH}}\right)\left(3785 \frac{\text{mL}}{\text{gal}}\right)$$

$$= 0.015 \text{ mL/d}$$

(C) This incorrect solution misuses the definition of pOH to find the hydroxide concentration. Other assumptions are unchanged from the correct solution.

$$\text{pH} + \text{pOH} = 14$$

At pH of 1.6,
$$pOH = 14 - 1.6 = 12.4$$
At pH of 7.0,
$$pOH = 14 - 7.0 = 7.0$$
$$OH^- \text{ required} = 10^{-12.4+7.0} = 10^{-5.4}$$
$$= 4.0 \times 10^{-6} \text{ mol/L}$$
$$NaOH \rightarrow Na^+ + OH^-$$

One mole of NaOH dissociates to produce one mole of OH^-.

$$\text{NaOH required} = 4.0 \times 10^{-6} \text{ mol/L}$$

The daily sodium hydroxide feed rate is

$$\frac{\left(50{,}000\,\dfrac{\text{gal waste}}{\text{day}}\right)\left(4.0 \times 10^{-6}\,\dfrac{\text{mol NaOH}}{\text{L waste}}\right)}{0.005\,\dfrac{\text{meq}}{\text{L acid}}} \times \left(1\,\dfrac{\text{equiv}}{\text{mol NaOH}}\right)\left(3785\,\dfrac{\text{mL}}{\text{gal}}\right)$$

$$= 1.5 \times 10^5 \text{ mL/d}$$

(D) This incorrect solution confuses the relationship between the acid requirement and its corresponding concentration. Other assumptions are unchanged from the correct solution.

$$pH + pOH = 14$$

At pH of 1.6,
$$pOH = 14 - 1.6 = 12.4$$
$$[OH^-] = 10^{-12.4} \text{ mol/L}$$
$$= 4.0 \times 10^{-13} \text{ mol/L}$$

At pH of 7.0,
$$pOH = 14 - 7.0 = 7.0$$
$$[OH^-] = 10^{-7.0}\,\dfrac{\text{mol}}{\text{L}} = 1.0 \times 10^{-7} \text{ mol/L}$$
$$OH^- \text{ required} = 1.0 \times 10^{-7}\,\dfrac{\text{mol}}{\text{L}} - 4.0 \times 10^{-13}\,\dfrac{\text{mol}}{\text{L}}$$
$$= 1.0 \times 10^{-7} \text{ mol/L}$$
$$NaOH \rightarrow Na^+ + OH^-$$

One mole of NaOH dissociates to produce one mole of OH^-.

$$\text{NaOH required} = 1.0 \times 10^{-7} \text{ mol/L}$$

The daily sodium hydroxide feed rate is

$$\frac{\left(50{,}000\,\dfrac{\text{gal}}{\text{day}}\right)\left(0.005\,\dfrac{\text{meq}}{\text{L}}\right)\left(3785\,\dfrac{\text{mL}}{\text{gal}}\right)}{\left(1.0 \times 10^{-7}\,\dfrac{\text{mol}}{\text{L}}\right)\left(1\,\dfrac{\text{meq}}{\text{mol}}\right)}$$

$$= 9.5 \times 10^{12} \text{ mL/d}$$

SOLUTION 71

The waste volume is

$$\frac{325{,}000\,\dfrac{\text{kg}}{\text{d}}}{98\,\dfrac{\text{kg}}{\text{m}^3}} = 3316 \text{ m}^3/\text{d}$$

The energy content is

$$\left(3316\,\dfrac{\text{m}^3}{\text{d}}\right)\left(890{,}000\,\dfrac{\text{kJ}}{\text{m}^3}\right) = 2.95 \times 10^9 \text{ kJ/d}$$

The bed area is

$$\frac{2.95 \times 10^9\,\dfrac{\text{kJ}}{\text{d}}}{\left(650{,}000\,\dfrac{\text{kJ}}{\text{m}^2\cdot\text{min}}\right)\left(1440\,\dfrac{\text{min}}{\text{d}}\right)} = 3.2 \text{ m}^2$$

The answer is (B).

Why Other Options Are Wrong

(A) This incorrect solution divides the discarded energy content by the loading rate without applying the waste generation rate.

The bed area is

$$\frac{890{,}000\,\dfrac{\text{kJ}}{\text{m}^3}}{650{,}000\,\dfrac{\text{kJ}}{\text{m}^2\cdot\text{min}}} = 1.4 \text{ m}^2$$

Units do not work.

(C) This incorrect solution improperly converts minutes to days.

The waste volume is

$$\frac{325{,}000\,\dfrac{\text{kg}}{\text{d}}}{98\,\dfrac{\text{kg}}{\text{m}^3}} = 3316 \text{ m}^3/\text{d}$$

The energy content is

$$\left(3316\,\dfrac{\text{m}^3}{\text{d}}\right)\left(890{,}000\,\dfrac{\text{kJ}}{\text{m}^3}\right) = 2.95 \times 10^9 \text{ kJ/d}$$

The bed area is

$$\frac{2.95 \times 10^9 \, \frac{\text{kJ}}{\text{d}}}{\left(650\,000 \, \frac{\text{kJ}}{\text{m}^2 \cdot \text{min}}\right)\left(60 \, \frac{\text{min}}{\text{d}}\right)} = 76 \text{ m}^2$$

(D) This incorrect solution calculates the waste volume by multiplying instead of dividing by the density.

The waste volume is

$$\left(325\,000 \, \frac{\text{kg}}{\text{d}}\right)\left(98 \, \frac{\text{kg}}{\text{m}^3}\right) = 3.2 \times 10^7 \text{ m}^3/\text{d}$$

Units do not work.

The energy content is

$$\left(3.2 \times 10^7 \, \frac{\text{m}^3}{\text{d}}\right)\left(8.9 \times 10^5 \, \frac{\text{kJ}}{\text{m}^3}\right) = 2.85 \times 10^{13} \text{ kJ/d}$$

The bed area is

$$\frac{2.85 \times 10^{13} \, \frac{\text{kJ}}{\text{d}}}{\left(650\,000 \, \frac{\text{kJ}}{\text{m}^2 \cdot \text{min}}\right)\left(1440 \, \frac{\text{min}}{\text{d}}\right)}$$

$$= 30\,427 \text{ m}^2 \quad (30\,000 \text{ m}^2)$$

SOLUTION 72

D_d	diffusion coefficient	m²/s
D^*	effective diffusion	m²/s
ω	tortuosity	–

$$D^* = \omega D_d = (0.6)\left(8.7 \times 10^{-9} \, \frac{\text{m}^2}{\text{s}}\right)$$

$$= 5.2 \times 10^{-9} \text{ m}^2/\text{s}$$

C	concentration of the solute at time t	mg/L
C_o	concentration of the solute at time zero	mg/L
erfc	complementary error function	–
t	travel time of interest	day
x	liner thickness	m

$$\frac{C}{C_o} = \text{erfc} \frac{x}{2\sqrt{D^*t}} = \frac{100 \, \frac{\text{mg}}{\text{L}}}{12\,000 \, \frac{\text{mg}}{\text{L}}} = 0.008\,33$$

Using complimentary error function tables,

$$\frac{x}{2\sqrt{D^*t}} = 1.87$$

$$x = (1.87)(2)\sqrt{\left(5.2 \times 10^{-9} \, \frac{\text{m}^2}{\text{s}}\right)(100 \text{ yr}) \times \left(365 \, \frac{\text{d}}{\text{yr}}\right)\left(86\,400 \, \frac{\text{s}}{\text{d}}\right)}$$

$$= 15 \text{ m}$$

The answer is (B).

Why Other Options Are Wrong

(A) This incorrect choice misuses the complementary error function. Other assumptions, definitions, and equations are the same as used in the correct solution.

$$D^* = (0.6)\left(8.7 \times 10^{-9} \, \frac{\text{m}^2}{\text{s}}\right)$$

$$= 5.2 \times 10^{-9} \text{ m}^2/\text{s}$$

$$\frac{C}{C_o} = \text{erfc} \frac{x}{2(D^*t)^{0.5}} = \frac{100 \, \frac{\text{mg}}{\text{L}}}{12\,000 \, \frac{\text{mg}}{\text{L}}}$$

$$= 0.008\,33$$

$$\frac{x}{2(D^*t)^{0.5}} = 0.99$$

$$x = (0.99)(2)\sqrt{\left(5.2 \times 10^{-9} \, \frac{\text{m}^2}{\text{s}}\right)(100 \text{ yr}) \times \left(365 \, \frac{\text{d}}{\text{yr}}\right)\left(86\,400 \, \frac{\text{s}}{\text{d}}\right)}$$

$$= 8 \text{ m}$$

(C) This incorrect choice uses the diffusion coefficient for the effective diffusion. Other assumptions, definitions, and equations are the same as used in the correct solution.

$$D^* = 8.7 \times 10^{-9} \text{ m}^2/\text{s}$$

$$\frac{C}{C_o} = \text{erfc} \frac{x}{2\sqrt{D^*t}} = \frac{100 \, \frac{\text{mg}}{\text{L}}}{12\,000 \, \frac{\text{mg}}{\text{L}}}$$

$$= 0.008\,33$$

$$\frac{x}{2\sqrt{D^*t}} = 1.87$$

$$x = (1.87)(2)\sqrt{\left(8.7 \times 10^{-9} \, \frac{\text{m}^2}{\text{s}}\right)(100 \text{ yr}) \times \left(365 \, \frac{\text{d}}{\text{yr}}\right)\left(86\,400 \, \frac{\text{s}}{\text{d}}\right)}$$

$$= 20 \text{ m}$$

(D) This incorrect choice neglects to take the square root of the effective diffusion-time term. Other assumptions, definitions, and equations are the same as used in the correct solution.

$$D^* = (0.6)\left(8.7 \times 10^{-9} \, \frac{\text{m}^2}{\text{s}}\right)$$

$$= 5.2 \times 10^{-9} \, \frac{\text{m}^2}{\text{s}}$$

$$\frac{C}{C_o} = \text{erfc} \frac{x}{2\sqrt{D^*t}} = \frac{100 \, \frac{\text{mg}}{\text{L}}}{12\,000 \, \frac{\text{mg}}{\text{L}}}$$

$$= 0.008\,33$$

$$\frac{x}{2\sqrt{D^*t}} = 1.87$$

$$x = (1.87)(2)\left(5.2 \times 10^{-9}\ \frac{\text{m}^2}{\text{s}}\right)$$
$$\times (100\ \text{yr})\left(365\ \frac{\text{d}}{\text{yr}}\right)\left(86\,400\ \frac{\text{s}}{\text{d}}\right)$$
$$= 61\ \text{m}$$

SOLUTION 73

Selecting the "no action" alternative does not mean the responsible party can simply walk away from the site. What it does mean is that direct action to remediate the contamination will not occur, but other activities will be implemented at the site. Ongoing monitoring of the groundwater will be required to ensure that conditions remain static—if mobility of the contaminant is detected, the no action alternative may have to be abandoned. Ongoing monitoring may involve periodic collection of groundwater samples and analysis for the contaminants and degradation products.

Limitations on current uses and certain kinds of future development of the site may be applied. For example, activities involving deep excavations where dewatering would be required may be restricted, as would any kind of pumping well. Agricultural uses involving heavy irrigation that would significantly influence the groundwater gradient or water-table elevation may be restricted. Deed restrictions would be placed on the property to convey any restricted land uses to subsequent owners. Because of these restrictions, it may not be economically desirable to apply the no action alternative at some sites that meet the technical requirements.

Under some conditions involving complicated soil/groundwater systems and contaminants that are exceptionally difficult to remediate, or where excessive costs may be incurred, a technically or economically feasible alternative to no direct action may not exist.

The answer is (D).

Why Other Options Are Wrong

(A) This choice is incorrect because the responsible party retains indefinite responsibility for the site and it is unlikely that any release from future responsibility would occur. However, sites where remediation is exceptionally difficult or excessive costs may be incurred are potential candidates for no action.

(B) This choice is incorrect because the responsible party is not released from any future responsibility for the site. However, ongoing site monitoring may be required to ensure that conditions remain static and deed or future use restrictions may be imposed on the property.

(C) This incorrect choice does not include all applicable statements. Although ongoing site monitoring may be required to ensure that conditions remain static and deed or future use restrictions may be imposed on such properties, included among these sites are those where remediation is exceptionally difficult or excessive costs may be incurred.

SOLUTION 74

Use a 100 kg sample so that dry mass can be conveniently expressed in kg.

component	dry mass (kg)	density (kg/m³)	volume (m³)	recyclables volume (m³)
paper and paper products	36	140	0.257	0.257
yard waste	18	120	0.150	–
food waste	9.0	300	0.030	–
ferrous metals	5.1	160	0.032	0.032
non-ferrous metals	4.3	240	0.018	0.018
plastics	7.6	130	0.058	0.058
glass	6.9	350	0.020	0.020
wood	3.9	220	0.018	–
textiles	2.1	60	0.035	–
rubber	3.2	130	0.025	–
miscellaneous inert materials	3.9	480	0.008	–
			0.651	0.385

The percentage of the total waste volume that can potentially be recycled is

$$\frac{(0.385\ \text{m}^3)(0.92) \times 100\%}{0.651\ \text{m}^3} = 54\%$$

The answer is (B).

Why Other Options Are Wrong

(A) This incorrect option calculates the percent of waste not recycled by the participating population. The table used in the correct solution is unchanged. The percentage of the total waste volume that can potentially be recycled is

$$\frac{(0.651\ \text{m}^3 - 0.385\ \text{m}^3) \times 100\%}{0.651\ \text{m}^3} = 41\%$$

(C) This incorrect option calculates the total population by the percent participating and bases volumes on this value.

$$(75\,000\ \text{people})(0.92)\left(1.9\ \frac{\text{kg}}{\text{person} \cdot \text{d}}\right) = 131\,100\ \text{kg/d}$$

component	dry mass (%)	dry mass (kg/d)	density (km/m³)	volume (m³/d)	recyclables volume (m³/d)
paper and paper products	36	47 196	140	337	337
yard waste	18	23 598	120	197	
food waste	9.0	11 799	300	39.3	
ferrous metals	5.1	6686	160	41.8	41.8
non-ferrous metals	4.3	5637	240	23.5	23.5
plastics	7.6	9964	130	76.6	76.6
glass	6.9	9046	351	25.8	25.8
wood	3.9	5113	220	23.2	
textiles	2.1	2753	60	45.9	
rubber	3.2	4195	130	32.3	
miscellaneous inert materials	3.9	5113	478	10.7	
				853	505

The percentage of the total waste volume that can potentially be recycled is

$$\frac{(505 \text{ m}^3) \times 100\%}{853 \text{ m}^3} = 59\%$$

(D) This incorrect option adjusts for the participating population by dividing by 92% instead of multiplying. The table used in the correct solution is unchanged. The percentage of the total waste volume that can potentially be recycled is

$$\frac{(0.385 \text{ m}^3) \times 100\%}{(0.651 \text{ m}^3)(0.92)} = 64\%$$

SOLUTION 75

The transfer station will become more economical than direct-haul when the station's cost—the sum of amortized capital, operating costs, and the corresponding haul costs—is less than the direct-haul cost. This occurs at the breakeven point where the two costs are equal.

C_{dh} direct-haul breakeven cost $/ton-min

$$\frac{\$3.87}{\text{ton}} + \left(\frac{\$0.016}{\text{ton-min}}\right)(72 \text{ min}) = (C_{dh})(72 \text{ min})$$

$$C_{dh} = \frac{\dfrac{\$3.87}{\text{ton}} + \dfrac{\$1.15}{\text{ton}}}{72 \text{ min}} = \$0.070/\text{ton-min}$$

The answer is (C).

Why Other Options Are Wrong

(A) This incorrect choice equates the sum of the transportation costs to the amortized transfer station costs. Other definitions are unchanged from the correct solution.

$$\frac{\$3.87}{\text{ton}} = (C_{dh})(72 \text{ min}) + \left(\frac{\$0.016}{\text{ton-min}}\right)(72 \text{ min})$$

$$C_{dh} = \frac{\dfrac{\$3.87}{\text{ton}} - \dfrac{\$1.15}{\text{ton}}}{72 \text{ min}} = \$0.038/\text{ton-min}$$

(B) This incorrect choice uses the cost for direct-haul instead of the cost of the transfer station, and then misapplies the result. Other assumptions and definitions are unchanged from the correct solution.

$$\frac{\$3.87}{\text{ton}} = \left(\frac{\$0.061}{\text{ton-min}}\right)(72 \text{ min}) + (C_{dh})(72 \text{ min})$$

$$C_{dh} = \frac{\dfrac{\$3.87}{\text{ton}} - \dfrac{\$4.39}{\text{ton}}}{72 \text{ min}} = -\$0.0072/\text{ton-min}$$

$$\frac{\$0.061}{\text{ton-min}} - \frac{\$0.0072}{\text{ton-min}} - \$0.054/\text{ton-min}$$

(D) This incorrect choice uses the cost for direct-haul instead of the transfer station haul cost in the calculation. Other assumptions and definitions are unchanged from the correct solution.

$$\frac{\$3.87}{\text{ton}} + \left(\frac{\$0.061}{\text{ton-min}}\right)(72 \text{ min}) = (C_{dh})(72 \text{ min})$$

$$C_{dh} = \frac{\dfrac{\$3.87}{\text{ton}} + \dfrac{\$4.39}{\text{ton}}}{72 \text{ min}} = \$0.11/\text{ton-min}$$

SOLUTION 76

R rating   –
WF weighting factor –
WR weighted rating –

$$WR = (WF)(R)$$

criteria	WF	site 1 R	site 1 WR	site 2 R	site 2 WR
haul distance	2	2	4	3	6
access routes	4	3	12	1	4
land value	3	2	6	2	6
permeability	3	3	9	4	12
heterogeneities	4	2	8	3	12
cover quantities	3	3	9	2	6
seismic activity	4	4	16	4	16
quality	2	3	6	3	6
gradient	1	2	2	3	3
depth	2	4	8	3	6
drainage pattern	3	3	9	2	6
streams	2	3	6	2	4
population	3	2	6	3	9
land uses	4	4	16	3	12
opposition	4	3	12	2	8
	44		129		116

WA   weighted average   –

$$WA = \frac{\sum WR}{\sum WF}$$

$$\text{WA site 1} = \frac{129}{44} = 2.9$$

$$\text{WA site 2} = \frac{116}{44} = 2.6$$

$$2.9 > 2.6$$

Choose site 1, for which WA = 2.9.

The answer is (C).

Why Other Options Are Wrong

(A) This incorrect choice divides the sum of the site ratings by the sum of the weighting factors. Other definitions are the same as in the correct solution.

criteria	WF	site 1 R	site 2 R
haul distance	2	2	3
access routes	4	3	1
land value	3	2	2
permeability	3	3	4
heterogeneities	4	2	3
cover quantities	3	3	2
seismic activity	4	4	4
quality	2	3	3
gradient	1	2	3
depth	2	4	3
drainage pattern	3	3	2
streams	2	3	2
population	3	2	3
land uses	4	4	3
opposition	4	3	2
	44	43	40

$$WA = \frac{\sum R}{\sum WF}$$

$$\text{WA site 1} = \frac{43}{44} = 0.98$$

$$\text{WA site 2} = \frac{40}{44} = 0.91$$

$$0.98 > 0.91$$

Choose site 1, for which WA = 0.98.

(B) This incorrect choice selects site 2 because it has the lower score. The sites are evaluated on a scale from 1 (unimportant/poor) to 4 (very important/excellent) so that the highest scoring site would be the preferred site. The table and other definitions and equations are the same as in the correct solution.

$$\text{WA site 1} = \frac{127}{44} = 2.9$$

$$\text{WA site 2} = \frac{116}{44} = 2.6$$

$$2.6 < 2.9$$

Choose site 2, for which WA = 2.6.

(D) This incorrect choice divides the sum of the weighted ratings by the number of criteria. The table and other definitions and equations are the same as in the correct solution.

n number of criteria

$$WA = \frac{\sum WR}{n}$$

$$\text{WA site 1} = \frac{129}{15} = 8.6$$

$$\text{WA site 2} = \frac{116}{15} = 7.7$$

$$8.6 > 7.7$$

Choose site 1, for which WA = 8.6.

SOLUTION 77

The dynamic viscosity and density for water at 10°C are 0.001 307 kg/m·s and 999.7 kg/m³, respectively.

g	gravitational constant	m/s²
i	groundwater gradient	–
k	intrinsic permeability	m²
n_e	soil effective porosity	–
v_x	groundwater velocity	m/d
μ_w	water dynamic viscosity	kg/m·s
ρ_w	water density	kg/m³

$$v_x = \frac{kg\rho_w i}{n_e \mu_w}$$

$$= \frac{(1.1 \times 10^{-5} \text{mm}^2)(9.81 \frac{\text{m}}{\text{s}^2}) \times (999.7 \frac{\text{kg}}{\text{m}^3})(0.00063)}{(0.38)(0.001\,307 \frac{\text{kg}}{\text{m·s}})}$$

$$\times \left(1000 \frac{\text{mm}}{\text{m}}\right)^2 \left(\frac{1 \text{ d}}{86\,400 \text{ s}}\right)$$

$$= 0.0118 \text{ m/d}$$

f_{oc}	organic carbon fraction	–
K_d	distribution coefficient	mL/g
K_{oc}	soil-water partition coefficient	mL/g

$$K_d = K_{oc} f_{oc} = \left(173 \, \frac{\text{mL}}{\text{g}}\right) \left(485 \, \frac{\text{mg}}{\text{kg}}\right) \left(\frac{1 \, \text{kg}}{10^6 \, \text{mg}}\right)$$
$$= 0.084 \, \text{mL/g}$$

B_d soil bulk density g/cm³
r_f retardation factor –

$$r_f = 1 + \frac{B_d K_d}{n_e}$$
$$= 1 + \frac{\left(1.8 \, \frac{\text{g}}{\text{cm}^3}\right) \left(0.084 \, \frac{\text{mL}}{\text{g}}\right) \left(1 \, \frac{\text{cm}^3}{\text{mL}}\right)}{0.38}$$
$$= 1.4$$

v_s dissolved organic solvent velocity m/d

$$v_s = \frac{v_x}{r_f} = \frac{0.0118 \, \frac{\text{m}}{\text{d}}}{1.4}$$
$$= 0.0084 \, \text{m/d}$$

The answer is (C).

Why Other Options Are Wrong

(A) This incorrect solution uses the intrinsic permeability as the hydraulic conductivity and the water-soil partition coefficient as the distribution coefficient. Other definitions and equations are unchanged from the correct solution.

K hydraulic conductivity m/d

$$v_x = \frac{Ki}{n_e} = \frac{\left(1.1 \times 10^{-5} \, \frac{\text{m}}{\text{d}}\right)(0.00063)}{0.38}$$
$$= 1.8 \times 10^{-8} \, \text{m/d}$$

This equation assumed the units for K to be m/d.

$$r_f = 1 + \frac{\left(1.8 \, \frac{\text{g}}{\text{cm}^3}\right) \left(173 \, \frac{\text{mL}}{\text{g}}\right) \left(1 \, \frac{\text{cm}^3}{\text{mL}}\right)}{0.38} = 820$$

$$v_s = \frac{1.8 \times 10^{-8} \, \frac{\text{m}}{\text{d}}}{820} = 2.2 \times 10^{-11} \, \text{m/d}$$

(B) This incorrect solution uses the Darcy velocity instead of the actual velocity for the groundwater and uses the partition coefficient as the distribution coefficient. Other definitions and equations are unchanged from the correct solution.

$$v_x = \frac{kg\rho_w i}{\mu_w}$$

$$= \frac{(1.1 \times 10^{-5} \, \text{mm}^2)\left(9.81 \, \frac{\text{m}}{\text{s}^2}\right)}{\left(0.001307 \, \frac{\text{kg}}{\text{m} \cdot \text{s}}\right)\left(1000 \, \frac{\text{mm}}{\text{m}}\right)^2 \left(\frac{1 \, \text{d}}{86400 \, \text{s}}\right)}$$

$$= 0.0045 \, \text{m/d}$$

$$r_f = 1 + \frac{\left(1.8 \, \frac{\text{g}}{\text{cm}^3}\right)\left(173 \, \frac{\text{mL}}{\text{g}}\right)\left(1 \, \frac{\text{cm}^3}{\text{mL}}\right)}{0.38} = 820$$

$$v_s = \frac{0.045 \, \frac{\text{m}}{\text{d}}}{820} = 5.5 \times 10^{-5} \, \text{m/d}$$

(D) This incorrect solution uses the groundwater velocity for the dissolved organic solvent velocity. Other definitions and equations are unchanged from the correct solution.

$$v_x = \frac{(1.1 \times 10^{-5} \, \text{mm}^2)\left(9.81 \, \frac{\text{m}}{\text{s}^2}\right) \times \left(999.7 \, \frac{\text{kg}}{\text{m}^3}\right)(0.00063)}{(0.38)\left(0.001307 \, \frac{\text{kg}}{\text{m} \cdot \text{s}}\right)\left(1000 \, \frac{\text{mm}}{\text{m}}\right)^2 \left(\frac{1 \, \text{d}}{86400 \, \text{s}}\right)}$$

$$= 0.0118 \, \text{m/d} \quad (0.012 \, \text{m/d})$$

SOLUTION 78

To reduce resin deterioration, the influent solution must be diluted to a maximum concentration of 11%.

C_1 undiluted concentration %
C_2 diluted concentration %
Q_D dilution flow rate m³/d
Q_w waste flow rate m³/d
V_D fractional dilution volume –

For the feed,

$$V_D = \frac{C_1}{C_2} - 1 = \frac{15\%}{11\%} - 1 = 0.36$$

To meet reuse specifications, the recovered solution must be diluted to 3.5 molar.

The mole weight of CrO_3 is

$$52 \, \frac{\text{g}}{\text{mol}} + (3)\left(16 \, \frac{\text{g}}{\text{mol}}\right) = 100 \, \text{g/mol}$$

$$Q_D = V_D Q_w = (0.36)\left(1000 \, \frac{\text{m}^3}{\text{d}}\right)$$
$$= 360 \, \text{m}^3/\text{d}$$

For the recovered solution, the equivalent percent concentration of 3.5 molar is

$$\left(3.5 \ \frac{\text{mol}}{\text{L}}\right)\left(100 \ \frac{\text{g}}{\text{mol}}\right)\left(\frac{1 \ \text{L}}{1000 \ \text{g}}\right) \times 100\% = 35\%$$

$$V_D = \frac{42\%}{35\%} - 1 = 0.20$$

$$Q_D = V_D Q_w = (0.20)\left(350 \ \frac{\text{m}^3}{\text{d}}\right)$$
$$= 70 \ \text{m}^3/\text{d}$$

For total dilution,

$$Q_D = 360 \ \frac{\text{m}^3}{\text{d}} + 70 \ \frac{\text{m}^3}{\text{d}} = 430 \ \text{m}^3/\text{d}$$

The answer is (A).

Why Other Options Are Wrong

(B) This incorrect choice bases the dilution requirement on the ratio of 15% to 3.5 molar. Other assumptions, definitions, and equations are the same as the correct solution.

The mole weight of CrO_3 is

$$52 \ \frac{\text{g}}{\text{mol}} + (3)\left(16 \ \frac{\text{g}}{\text{mol}}\right) = 100 \ \text{g/mol}$$

The equivalent percent concentration of 3.5 molar is

$$\left(3.5 \ \frac{\text{mol}}{\text{L}}\right)\left(100 \ \frac{\text{g}}{\text{mol}}\right)\left(\frac{1 \ \text{L}}{1000 \ \text{g}}\right) \times 100\% = 35\%$$

$$V_D = \frac{35\%}{15\%} - 1 = 1.3$$

$$(1.3)\left(1000 \ \frac{\text{m}^3}{\text{d}}\right) = 1300 \ \text{m}^3/\text{d}$$

(C) This incorrect choice defines the fractional dilution volume as the ratio of the undiluted to the diluted concentrations. Other assumptions, definitions, and equations are the same as the correct solution. For the feed,

$$V_D = \frac{C_1}{C_2} = \frac{15\%}{11\%} = 1.4$$

$$Q_D = (1.4)\left(1000 \ \frac{\text{m}^3}{\text{d}}\right) = 1400 \ \text{m}^3/\text{d}$$

For the recovered solution, the mole weight of CrO_3 is

$$52 \ \frac{\text{g}}{\text{mol}} + (3)\left(16 \ \frac{\text{g}}{\text{mol}}\right) = 100 \ \text{g/mol}$$

The equivalent percent concentration of 3.5 molar is

$$\left(3.5 \ \frac{\text{mol}}{\text{L}}\right)\left(100 \ \frac{\text{g}}{\text{mol}}\right)\left(\frac{1 \ \text{L}}{1000 \ \text{g}}\right) \times 100\% = 35\%$$

$$V_D = \frac{42\%}{35\%} = 1.2$$

$$Q_D = (1.2)\left(350 \ \frac{\text{m}^3}{\text{d}}\right) = 420 \ \text{m}^3/\text{d}$$

For total dilution,

$$Q_D = 1400 \ \frac{\text{m}^3}{\text{D}} + 420 \ \frac{\text{m}^3}{\text{d}}$$
$$= 1820 \ \text{m}^3/\text{d} \quad (1800 \ \text{m}^3/\text{d})$$

(D) This incorrect choice bases the dilution requirement on the ratio of 11% to 3.5 molar. Other assumptions, definitions, and equations are the same as the correct solution.

The mole weight of CrO_3 is

$$52 \ \frac{\text{g}}{\text{mol}} + (3)\left(16 \ \frac{\text{g}}{\text{mol}}\right) = 100 \ \text{g/mol}$$

The equivalent percent concentration of 3.5 molar is

$$\left(3.5 \ \frac{\text{mol}}{\text{L}}\right)\left(100 \ \frac{\text{g}}{\text{mol}}\right)\left(\frac{1 \ \text{L}}{1000 \ \text{g}}\right) \times 100\% = 35\%$$

$$V_D = \frac{35\%}{11\%} - 1 = 2.2$$

$$(2.2)\left(1000 \ \frac{\text{m}^3}{\text{d}}\right) = 2200 \ \text{m}^3/\text{d}$$

SOLUTION 79

The constituents of interest are chromium trioxide (CrO_3), chromium sulfate ($Cr_2(SO_4)_3$), and chromium hydroxide ($Cr(OH)_3$).

The mole weight of CrO_3 is

$$52 \ \frac{\text{g}}{\text{mol}} + (3)\left(16 \ \frac{\text{g}}{\text{mol}}\right) = 100 \ \text{g/mol}$$

The molar concentration of CrO_3 is

$$\frac{165 \ \frac{\text{mg CrO}_3}{\text{L}}}{\left(100 \ \frac{\text{g}}{\text{mol}}\right)\left(1000 \ \frac{\text{mg}}{\text{g}}\right)} = 0.00165 \ \text{mol CrO}_3/\text{L}$$

The mole weight of $Cr(OH)_3$ is

$$52 \ \frac{\text{g}}{\text{mol}} + (3)\left(16 \ \frac{\text{g}}{\text{mol}} + 1 \ \frac{\text{g}}{\text{mol}}\right) = 103 \ \text{g/mol}$$

One mole of CrO_3 will react to produce 2/4 mol of $Cr_2(SO_4)_3$. The molar concentration of $Cr_2(SO_4)_3$ is

$$\left(0.00165 \ \frac{\text{mol CrO}_3}{\text{L}}\right)\left(\frac{\frac{2}{4} \ \text{mol Cr}_2(SO_4)_3}{1 \ \text{mol CrO}_3}\right)$$
$$= 0.000825 \ \text{mol Cr}_2(SO_4)_3/\text{L}$$

One mole of $Cr_2(SO_4)_3$ will react to produce 2 mol of $Cr(OH)_3$ sludge

$$\left(0.000825 \ \frac{\text{mol Cr}_2(SO_4)_3}{\text{L}}\right)\left(\frac{2 \ \text{mol Cr(OH)}_3}{1 \ \text{mol Cr}_2(SO_4)_3}\right)$$
$$= 0.00165 \ \text{molCr(OH)}_2/\text{L}$$

The mass of sludge produced is

$$\left(0.00165 \; \frac{\text{mol Cr(OH)}_3}{\text{L}}\right) \left(103 \; \frac{\text{g}}{\text{mol}}\right) \left(50{,}000 \; \frac{\text{gal}}{\text{day}}\right)$$
$$\times \left(3.785 \; \frac{\text{L}}{\text{gal}}\right) \left(\frac{1 \text{ kg}}{1000 \text{ g}}\right)$$
$$= 32 \text{ kg/d}$$

The answer is (A).

Why Other Options Are Wrong

(B) This incorrect solution miscalculates the molecular weight of chromium hydroxide. Other assumptions and definitions are unchanged from the correct solution.

The mole weight of CrO_3 is

$$52 \; \frac{\text{g}}{\text{mol}} + 3\left(16 \; \frac{\text{g}}{\text{mol}}\right) = 100 \text{ g/mol}$$

The molar concentration of CrO_3 is

$$\frac{165 \; \frac{\text{mg CrO}_3}{\text{L}}}{\left(100 \; \frac{\text{g}}{\text{mol}}\right)\left(1000 \; \frac{\text{mg}}{\text{g}}\right)} = 0.00165 \text{ mol CrO}_3/\text{L}$$

The mole weight of $Cr(OH)_3$ is

$$(3)\left(52 \; \frac{\text{g}}{\text{mol}} + 16 \; \frac{\text{g}}{\text{mol}} + 1 \; \frac{\text{g}}{\text{mol}}\right) = 207 \text{ g/mol}$$

One mole of CrO_3 will react to produce 2/4 mol of $Cr_2(SO_4)_3$. The molar concentration of $Cr_2(SO_4)_3$ is

$$\left(0.00165 \; \frac{\text{mol CrO}_3}{\text{L}}\right)\left(\frac{\frac{2}{4} \text{ mol Cr}_2(\text{SO}_4)_3}{1 \text{ mol CrO}_3}\right)$$
$$= 0.000825 \text{ mol Cr}_2(\text{SO}_4)_3/\text{L}$$

One mole of $Cr_2(SO_4)_3$ will react to produce 2 mol of $Cr(OH)_3$ sludge.

$$\left(0.000825 \; \frac{\text{mol Cr}_2(\text{SO}_4)_3}{\text{L}}\right)\left(\frac{2 \text{ mol Cr(OH)}_3}{1 \text{ mol Cr}_2(\text{SO}_4)_3}\right)$$
$$= 0.00165 \text{ mol Cr(OH)}_2/\text{L}$$

The mass of sludge produced is

$$\left(0.00165 \; \frac{\text{mol Cr(OH)}_3}{\text{L}}\right)\left(207 \; \frac{\text{g}}{\text{mol}}\right)\left(50{,}000 \; \frac{\text{gal}}{\text{day}}\right)$$
$$\times \left(3.785 \; \frac{\text{L}}{\text{gal}}\right)\left(\frac{1 \text{ kg}}{1000 \text{ g}}\right)$$
$$= 65 \text{ kg/d}$$

(C) This incorrect solution uses the molecular weight of chromium sulfate instead of chromium hydroxide. Other assumptions and definitions are unchanged from the correct solution.

The mole weight of CrO_3 is

$$52 \; \frac{\text{g}}{\text{mol}} + (3)\left(16 \; \frac{\text{g}}{\text{mol}}\right) = 100 \text{ g/mol}$$

The molar concentration of CrO_3 is

$$\frac{165 \; \frac{\text{mg CrO}_3}{\text{L}}}{\left(100 \; \frac{\text{g}}{\text{mol}}\right)\left(1000 \; \frac{\text{mg}}{\text{g}}\right)} = 0.00165 \text{ mol CrO}_3/\text{L}$$

The mole weight of $Cr_2(SO_4)_3$ is

$$(2)\left(52 \; \frac{\text{g}}{\text{mol}}\right) + (3)\left(32 \; \frac{\text{g}}{\text{mol}} + 4\left(16 \; \frac{\text{g}}{\text{mol}}\right)\right)$$
$$= 392 \text{ g/mol}$$

One mole of CrO_3 will react to produce 2/4 mol of $Cr_2(SO_4)_3$. The molar concentration of $Cr_2(SO_4)_3$ is

$$\left(0.00165 \; \frac{\text{mol CrO}_3}{\text{L}}\right)\left(\frac{\frac{2}{4} \text{ mol Cr}_2(\text{SO}_4)_3}{1 \text{ mol CrO}_3}\right)$$
$$= 0.000825 \text{ mol Cr}_2(\text{SO}_4)_3/\text{L}$$

One mole of $Cr_2(SO_4)_3$ will react to produce 2 mol of $Cr(OH)_3$ sludge.

$$\left(0.000825 \; \frac{\text{mol Cr}_2(\text{SO}_4)_3}{\text{L}}\right)$$
$$\times \left(\frac{2 \text{ mol Cr(OH)}_3}{1 \text{ mol Cr}_2(\text{SO}_4)_3}\right)$$
$$= 0.00165 \text{ mol Cr(OH)}_3/\text{L}$$

The mass of sludge produced is

$$\left(0.00165 \; \frac{\text{mol Cr(OH)}_3}{\text{L}}\right)\left(392 \; \frac{\text{g}}{\text{mol}}\right)\left(50{,}000 \; \frac{\text{gal}}{\text{day}}\right)$$
$$\times \left(3.785 \; \frac{\text{L}}{\text{gal}}\right)\left(\frac{1 \text{ kg}}{1000 \text{ g}}\right)$$
$$= 122 \text{ kg/d} \quad (120 \text{ kg/d})$$

(D) This incorrect solution calculates sludge production using mass ratios instead of mole ratios. Other assumptions and definitions are unchanged from the correct solution.

One gram of CrO_3 will react to produce 2/4 g of $Cr_2(SO_4)_3$.

$$\left(165 \; \frac{\text{g CrO}_3}{\text{L}}\right)\left(\frac{\frac{2}{4} \text{ g Cr}_2(\text{SO}_4)_3}{1 \text{ g CrO}_3}\right)$$
$$= 82.5 \text{ g Cr}_2(\text{SO}_4)_3/\text{L}$$

One gram of $Cr_2(SO_4)_3$ will react to produce 2 g of $Cr(OH)_3$ sludge.

$$\left(82.5 \, \frac{g \, Cr_2(SO_4)_3}{L}\right)\left(\frac{2 \, g \, Cr(OH)_3}{1 \, g \, Cr_2(SO_4)_3}\right)$$
$$= 165 \, g \, Cr(OH)_3/L$$

The mass of sludge produced is

$$\left(165 \, \frac{g \, Cr(OH)_3}{L}\right)\left(50{,}000 \, \frac{gal}{day}\right)$$
$$\times \left(3.785 \, \frac{L}{gal}\right)\left(\frac{1 \, kg}{1000 \, g}\right)$$
$$= 31\,226 \, kg/d \quad (31\,000 \, kg/d)$$

SOLUTION 80

The time required per month to collect the current roll-off boxes, based on a 90 min round trip for each box, is

$$(34 \, locations)\left(3 \, \frac{boxes}{location}\right)\left(90 \, \frac{min}{round \, trip\text{-}box}\right)$$
$$\times \left(2 \, \frac{round \, trips}{mo}\right)\left(\frac{1 \, hr}{60 \, min}\right)$$
$$= 306 \, hr/mo$$

The dumpsters and compaction trucks will collect the same volume of waste on the same schedule, but the compaction trucks will haul an equivalent uncompacted volume of

$$(12 \, yd^3 \, compacted)\left(\frac{3 \, yd^3 \, uncompacted}{1 \, yd^3 \, compacted}\right) = 36 \, yd^3$$

To provide the same dumpster capacity as roll-off box capacity, each location will have

$$\frac{\left(18 \, \frac{yd^3}{box}\right)\left(3 \, \frac{boxes}{location}\right)}{6 \, \frac{yd^3}{dumpster}} = 9 \, dumpsters/location$$

The total dumpsters emptied per load will be

$$\frac{36 \, \frac{yd^3}{load}}{6 \, \frac{yd^3}{dumpster}} = 6 \, dumpsters/load$$

For each location, one truck will complete one round trip to service six dumpsters.

$$(34 \, locations)\left(2 \, \frac{round \, trips}{mo}\right)$$
$$\times \left(90 \, \frac{min}{round \, trip}\right)\left(\frac{1 \, hr}{60 \, min}\right)$$
$$= 102 \, hr/mo$$

Another truck will complete one round trip with one trip between locations to service three dumpsters at that location and three dumpsters at another location.

$$(34 \, locations)\left(\frac{2 \, round \, trips}{2 \, locations}\right)$$
$$\times \left(\frac{90 \, min}{round \, trip} + \frac{27 \, min}{round \, trip}\right)\left(\frac{1 \, hr}{60 \, min}\right)$$
$$= 66 \, hr$$

The total time per month using the proposed dumpsters and compaction trucks will be

$$102 \, \frac{hr}{mo} + 66 \, \frac{hr}{mo} = 168 \, hr/mo$$

The total driving time saved is

$$306 \, \frac{hr}{mo} - 168 \, \frac{hr}{mo} = 138 \, hr/mo \quad (140 \, hr/mo)$$

The answer is (D).

Why Other Options Are Wrong

(A) This incorrect solution includes only one instead of three roll-off boxes per location.

The time required per month using the current roll-off boxes, based on a 90 min round trip for each box, is

$$(34 \, boxes)\left(\frac{90 \, min}{round \, trip\text{-}box}\right)$$
$$\times \left(2 \, \frac{round \, trips}{mo}\right)\left(\frac{1 \, hr}{60 \, min}\right)$$
$$= 102 \, hr/mo$$

The dumpsters and compaction trucks will collect the same volume of waste on the same schedule, but the compaction trucks will haul an equivalent uncompacted volume of

$$(12 \, yd^3 \, compacted)\left(\frac{3 \, yd^3 \, uncompacted}{1 \, yd^3 \, compacted}\right) = 36 \, yd^3$$

To provide the same dumpster capacity as roll-off box capacity, each location will have

$$\left(\frac{18 \, \frac{yd^3}{box}}{6 \, \frac{yd^3}{dumpster}}\right)\left(1 \, \frac{box}{location}\right) = 3 \, dumpsters/location$$

The total dumpsters emptied per load will be

$$\frac{36 \, \frac{yd^3}{load}}{6 \, \frac{yd^3}{dumpster}} = 6 \, dumpsters/load$$

For each location, one truck will complete one round trip with one trip between locations to service three dumpsters at each of two locations. The total time per month using the proposed dumpsters and compaction trucks will be

$$(34 \text{ locations}) \left(\frac{2 \text{ round trips}}{2 \text{ locations}} \right)$$
$$\times \left(\frac{90 \text{ min}}{\text{round trip}} + \frac{27 \text{ min}}{\text{round trip}} \right) \left(\frac{1 \text{ hr}}{60 \text{ min}} \right)$$
$$= 66 \text{ hr}$$

The total driving time saved is

$$102 \frac{\text{hr}}{\text{mo}} - 66 \frac{\text{hr}}{\text{mo}} = 36 \text{ hr/mo}$$

(B) This incorrect solution considers the time for a one-way instead of a two-way trip.

The time required per month using the current roll-off boxes, based on a 45 minute trip for each box, is

$$(34 \text{ locations}) \left(\frac{3 \text{ boxes}}{\text{location}} \right) \left(\frac{45 \text{ min}}{\text{trip-box}} \right)$$
$$\times \left(2 \frac{\text{round trips}}{\text{mo}} \right) \left(\frac{1 \text{ hr}}{60 \text{ min}} \right)$$
$$= 153 \text{ hr}$$

The dumpsters and compaction trucks will collect the same volume of waste on the same schedule, but the compaction trucks will haul an equivalent uncompacted volume of

$$(12 \text{ yd}^3 \text{ compacted}) \left(\frac{3 \text{ yd}^3 \text{ uncompacted}}{1 \text{ yd}^3 \text{ compacted}} \right) = 36 \text{ yd}^3$$

To provide the same dumpster capacity as roll-off box capacity, each location will have

$$\frac{\left(\frac{18 \text{ yd}^3}{\text{box}} \right) \left(3 \frac{\text{boxes}}{\text{location}} \right)}{6 \frac{\text{yd}^3}{\text{dumpster}}} = 9 \text{ dumpsters/location}$$

The total dumpsters emptied per load will be

$$\frac{36 \frac{\text{yd}^3}{\text{load}}}{6 \frac{\text{yd}^3}{\text{dumpster}}} = 6 \text{ dumpsters/load}$$

For each location, one truck will complete one trip to service six dumpsters, and another truck will complete one trip with one trip between locations to service three dumpsters at that location and three dumpsters at another location.

$$(34 \text{ locations}) \left(\frac{2 \text{ round trips}}{\text{mo}} \right)$$
$$\times \left(\frac{45 \text{ min}}{\text{round trip}} \right) \left(\frac{1 \text{ hr}}{60 \text{ min}} \right)$$
$$= 51 \text{ hr}$$

$$(34 \text{ locations}) \left(\frac{\frac{2 \text{ round trips}}{\text{mo}}}{2 \text{ locations}} \right)$$
$$\times \left(\frac{45 \text{ min}}{\text{round trip}} + \frac{27 \text{ min}}{\text{round trip}} \right) \left(\frac{1 \text{ hr}}{60 \text{ min}} \right)$$
$$= 41 \text{ hr}$$

The total time per month using the proposed dumpsters and compaction trucks will be

$$51 \frac{\text{hr}}{\text{mo}} + 41 \frac{\text{hr}}{\text{mo}} = 92 \text{ hr/mo}$$

The total driving time saved is

$$153 \frac{\text{hr}}{\text{mo}} - 92 \frac{\text{hr}}{\text{mo}} = 61 \text{ hr/mo}$$

(C) This incorrect solution includes only one pick-up per month instead of the scheduled two.

The time required per month using the current roll-off boxes, based on a 90 min round trip for each box, is

$$(34 \text{ locations}) \left(\frac{3 \text{ boxes}}{\text{location}} \right) \left(\frac{90 \text{ min}}{\text{round trip-box}} \right) \left(\frac{1 \text{ hr}}{60 \text{ min}} \right)$$
$$= 153 \text{ hr}$$

The dumpsters and compaction trucks will collect the same volume of waste on the same schedule, but the compaction trucks will haul an equivalent uncompacted volume of

$$(12 \text{ yd}^3 \text{ compacted}) \left(\frac{3 \text{ yd}^3 \text{ uncompacted}}{1 \text{ yd}^3 \text{ compacted}} \right) = 36 \text{ yd}^3$$

To provide the same dumpster capacity as roll-off box capacity, each location will have

$$\frac{\left(\frac{18 \text{ yd}^3}{\text{box}} \right) \left(\frac{3 \text{ boxes}}{\text{location}} \right)}{6 \frac{\text{yd}^3}{\text{dumpster}}} = 9 \text{ dumpsters/location}$$

The total dumpsters emptied per load will be

$$\frac{36 \frac{\text{yd}^3}{\text{load}}}{6 \frac{\text{yd}^3}{\text{dumpster}}} = 6 \text{ dumpsters/load}$$

For each location, one truck will complete one round trip to service six dumpsters, and another truck will complete one round trip with one trip between locations to service three dumpsters at that location and three at another location.

$$(34 \text{ locations}) \left(1 \frac{\text{round trip}}{\text{mo}}\right)$$
$$\times \left(\frac{90 \text{ min}}{\text{round trip}}\right)\left(\frac{1 \text{ hr}}{60 \text{ min}}\right)$$
$$= 51 \text{ hr/mo}$$

$$(34 \text{ locations}) \left(\frac{1 \frac{\text{round trip}}{\text{mo}}}{2 \text{ locations}}\right)$$
$$\times \left(\frac{90 \text{ min}}{\text{round trip}} + \frac{27 \text{ min}}{\text{round trip}}\right)\left(\frac{1 \text{ hr}}{60 \text{ min}}\right)$$
$$= 33 \text{ hr}$$

The total time per month using the proposed dumpsters and compaction trucks will be

$$51 \frac{\text{hr}}{\text{mo}} + 33 \frac{\text{hr}}{\text{mo}} = 84 \text{ hr/mo}$$

The total driving time saved is

$$153 \frac{\text{hr}}{\text{mo}} - 84 \frac{\text{hr}}{\text{mo}} = 69 \text{ hr/mo}$$

SOLUTION 81

The natural or blended gas required is

$$\frac{\left(38,000 \frac{\text{ft}^3}{\text{day}}\right)(0.65)}{0.70} = 35,286 \text{ ft}^3/\text{day}$$

The current natural gas cost is

$$\left(35,286 \frac{\text{ft}^3}{\text{day}}\right)\left(\frac{\$0.020}{\text{ft}^3}\right) = \$706/\text{day}$$

The blended natural gas required is

$$\left(35,286 \frac{\text{ft}^3}{\text{day}}\right)(0.30) = 10,586 \text{ ft}^3/\text{day}$$

The blended natural gas cost is

$$\left(10,586 \frac{\text{ft}^3}{\text{day}}\right)\left(\frac{\$0.020}{\text{ft}^3}\right) = \$212/\text{day}$$

The scrubber cost is

$$\left(38,000 \frac{\text{ft}^3}{\text{day}}\right)\left(\frac{\$0.0080}{\text{ft}^3}\right) = \$304/\text{day}$$

The total cost for blended gas is

$$\frac{\$212}{\text{day}} + \frac{\$304}{\text{day}} = \$516/\text{day}$$

Assume the gas blending continues full time for 365 day/yr.

The annual savings realized is

$$\left(\frac{\$706}{\text{day}} - \frac{\$516}{\text{day}}\right)\left(365 \frac{\text{day}}{\text{yr}}\right)$$
$$= \$69,350/\text{yr} \quad (\$69,000/\text{yr})$$

The answer is (A).

Why Other Options Are Wrong

(B) This incorrect solution bases scrubber cost on methane gas recovered instead of digester gas scrubbed. Other assumptions are unchanged from the correct solution.

The natural or blended gas required is

$$\frac{\left(38,000 \frac{\text{ft}^3}{\text{day}}\right)(0.65)}{0.70} = 35,286 \text{ ft}^3/\text{day}$$

The current natural gas cost is

$$\left(35,286 \frac{\text{ft}^3}{\text{day}}\right)\left(\frac{\$0.020}{\text{ft}^3}\right) = \$706/\text{day}$$

The blended natural gas required is

$$\left(35,286 \frac{\text{ft}^3}{\text{day}}\right)(0.30) = 10,586 \text{ ft}^3/\text{day}$$

The blended natural gas cost is

$$\left(10,586 \frac{\text{ft}^3}{\text{day}}\right)\left(\frac{\$0.020}{\text{ft}^3}\right) = \$212/\text{day}$$

The scrubber cost is

$$\left(38,000 \frac{\text{ft}^3}{\text{day}}\right)(0.65)\left(\frac{\$0.0080}{\text{ft}^3}\right) = \$198/\text{day}$$

The total cost for blended gas is

$$\frac{\$212}{\text{day}} + \frac{\$198}{\text{day}} = \frac{\$410}{\text{day}}$$

The annual savings realized is

$$\left(\frac{\$706}{\text{day}} - \frac{\$410}{\text{day}}\right)\left(365 \frac{\text{day}}{\text{yr}}\right)$$
$$= \$108,040/\text{yr} \quad (\$110,000/\text{yr})$$

(C) This incorrect solution considers the digester gas mixture instead of only the methane gas fraction. Other assumptions are unchanged from the correct solution.

The natural or blended gas required is

$$\frac{38{,}000 \ \frac{\text{ft}^3}{\text{day}}}{0.70} = 54{,}286 \ \text{ft}^3/\text{day}$$

The current natural gas cost is

$$\left(54{,}286 \ \frac{\text{ft}^3}{\text{day}}\right)\left(\frac{\$0.020}{\text{ft}^3}\right) = \$1086/\text{day}$$

The blended natural gas required is

$$\left(54{,}286 \ \frac{\text{ft}^3}{\text{day}}\right)(0.30) = 16{,}286 \ \text{ft}^3/\text{day}$$

The blended natural gas cost is

$$\left(16{,}286 \ \frac{\text{ft}^3}{\text{day}}\right)\left(\frac{\$0.020}{\text{ft}^3}\right) = \$326/\text{day}$$

The scrubber cost is

$$\left(38{,}000 \ \frac{\text{ft}^3}{\text{day}}\right)\left(\frac{\$0.0080}{\text{ft}^3}\right) = \$304/\text{day}$$

The total cost for blended gas is

$$\frac{\$326}{\text{day}} + \frac{\$304}{\text{day}} = \$630/\text{day}$$

The annual savings realized is

$$\left(\frac{\$1086}{\text{day}} - \frac{\$630}{\text{day}}\right)\left(365 \ \frac{\text{day}}{\text{yr}}\right)$$
$$= \$166{,}440/\text{yr} \quad (\$170{,}000/\text{yr})$$

(D) This incorrect solution confuses the methane and natural gas blending fractions. Other assumptions are unchanged from the correct solution.

The natural or blended gas required is

$$\frac{\left(38{,}000 \ \frac{\text{ft}^3}{\text{day}}\right)(0.65)}{0.30} = 82{,}333 \ \text{ft}^3/\text{day}$$

The current natural gas cost is

$$\left(82{,}333 \ \frac{\text{ft}^3}{\text{day}}\right)\left(\frac{\$0.020}{\text{ft}^3}\right) = \$1647/\text{day}$$

The blended natural gas required is

$$\left(82{,}333 \ \frac{\text{ft}^3}{\text{day}}\right)(0.30) = 24{,}700 \ \text{ft}^3/\text{day}$$

The blended natural gas cost is

$$\left(24{,}700 \ \frac{\text{ft}^3}{\text{day}}\right)\left(\frac{\$0.020}{\text{ft}^3}\right) = \$494/\text{day}$$

The scrubber cost is

$$\left(38{,}000 \ \frac{\text{ft}^3}{\text{day}}\right)\left(\frac{\$0.0080}{\text{ft}^3}\right) = \$304/\text{day}$$

The total cost for blended gas is

$$\frac{\$494}{\text{day}} + \frac{\$304}{\text{day}} = \$798/\text{day}$$

The annual savings realized is

$$\left(\frac{\$1647}{\text{day}} - \frac{\$798}{\text{day}}\right)\left(365 \ \frac{\text{day}}{\text{yr}}\right)$$
$$= \$309{,}885/\text{yr} \quad (\$310{,}000/\text{yr})$$

GROUNDWATER AND WELL FIELDS

SOLUTION 82

ion	concentration (mg/L)	valence (equiv/mol)	mole weight (g/mol)	equivalent weight (mg/meq)	equivalent concentration (meq/L)
Ca^{2+}	128	2	40	20	6.4
Mg^{2+}	66	2	24	12	5.5
SO_4^{2-}	83	2	96	48	1.7
Cl^-	21	1	35	35	0.60
NO_3^-	14	1	62	62	0.23
HCO_3^-	279	1	61	61	4.6
Na^+	7	1	23	23	0.30

$$\sum \text{cations} = 6.4 \ \frac{\text{meq}}{\text{L}} + 5.5 \ \frac{\text{meq}}{\text{L}} + 0.3 \ \frac{\text{meq}}{\text{L}}$$
$$= 12.2 \ \text{meq/L}$$

$$\sum \text{anions} = 1.7 \ \frac{\text{meq}}{\text{L}} + 0.60 \ \frac{\text{meq}}{\text{L}}$$
$$+ 0.23 \ \frac{\text{meq}}{\text{L}} + 4.6 \ \frac{\text{meq}}{\text{L}}$$
$$= 7.13 \ \text{meq/L}$$

$$\sum \text{cations} \gg \sum \text{anions}$$

Not all ions that are likely present at a significant concentration in the sample are included in the analysis. It is deficient in anions.

The answer is (C).

Why Other Options Are Wrong

(A) This is an incorrect choice because the sum of the cations and the sum of the anions in equivalent concentration units are not balanced (nearly equal to each other). Selecting this option would either be a guess or require a mathematical error in calculating the equivalent concentrations.

(B) This incorrect choice results if the concentrations are summed for cations and anions in mass concentration units and the erroneous assumption is made that cations should be greater than anions, possibly thinking that the cations are more common.

$$\sum \text{cations} = 128 \ \frac{\text{mg}}{\text{L}} + 66 \ \frac{\text{mg}}{\text{L}} + 7 \ \frac{\text{mg}}{\text{L}}$$
$$= 201 \ \text{mg/L}$$

$$\sum \text{anions} = 83 \ \frac{\text{mg}}{\text{L}} + 21 \ \frac{\text{mg}}{\text{L}} + 14 \ \frac{\text{mg}}{\text{L}} + 279 \ \frac{\text{mg}}{\text{L}}$$
$$= 397 \ \text{mg/L}$$

$$\sum \text{anions} \gg \sum \text{cations}$$

(D) This incorrect choice results if the concentrations are summed for cations and anions in mass concentration units.

$$\sum \text{cations} = 128 \ \frac{\text{mg}}{\text{L}} + 66 \ \frac{\text{mg}}{\text{L}} + 7 \ \frac{\text{mg}}{\text{L}}$$
$$= 201 \ \text{mg/L}$$

$$\sum \text{anions} = 83 \ \text{mg} L + 21 \ \frac{\text{mg}}{\text{L}} + 14 \ \frac{\text{mg}}{\text{L}} + 279 \ \frac{\text{mg}}{\text{L}}$$
$$= 397 \ \text{mg/L}$$

$$\sum \text{cations} \gg \sum \text{anions}$$

SOLUTION 83

g	gravitational constant	32.2 ft/sec^2
k	intrinsic permeability	in^2
K	hydraulic conductivity	ft/sec
v	kinematic viscosity of water at 45°F	1.537×10^{-5} ft^2/sec

$$K = \frac{gk}{v}$$

$$= \frac{\left(32.2 \ \frac{\text{ft}}{\text{sec}^2}\right)(1.4 \times 10^{-6} \ \text{in}^2)\left(\frac{1 \ \text{ft}^2}{144 \ \text{in}^2}\right)}{1.537 \times 10^{-5} \ \frac{\text{ft}^2}{\text{sec}}}$$

$$= 0.0204 \ \text{ft/sec}$$

i	gradient	–
n_e	effective porosity	–
v_x	rate of advection	ft/day

$$v_x = \frac{Ki}{n_e}$$

$$= \frac{\left(0.0204 \ \frac{\text{ft}}{\text{sec}}\right)\left(86{,}400 \ \frac{\text{sec}}{\text{day}}\right)(0.000\,35)}{0.42}$$

$$= 1.5 \ \text{ft/day}$$

The answer is (B).

Why Other Options Are Wrong

(A) This incorrect solution fails to divide by effective porosity. The result is the Darcy velocity, not the rate of advection (actual velocity). Other assumptions, definitions, and equations are unchanged from the correct solution.

$$K = \frac{\left(32.2 \ \frac{\text{ft}}{\text{sec}^2}\right)(1.4 \times 10^{-6} \ \text{in}^2)\left(\frac{\text{ft}^2}{144 \ \text{in}^2}\right)}{1.537 \times 10^{-5} \ \frac{\text{ft}^2}{\text{sec}}}$$

$$= 0.0204 \ \text{ft/sec}$$

$$v_x = Ki = \left(0.0204 \ \frac{\text{ft}}{\text{sec}}\right)\left(86{,}400 \ \frac{\text{sec}}{\text{day}}\right)(0.000\,35)$$

$$= 0.62 \ \text{ft/day}$$

(C) This incorrect solution uses the viscosity value for SI units at 45°C, but applies the English units to the value and fails to divide by the effective porosity. Other assumptions, definitions, and equations are unchanged from the correct solution.

$$v = 6.086 \times 10^{-7} \ \text{ft}^2/\text{sec}$$

$$K = \frac{\left(32.2 \ \frac{\text{ft}}{\text{sec}^2}\right)(1.4 \times 10^{-6} \ \text{in}^2)\left(\frac{\text{ft}^2}{144 \ \text{in}^2}\right)}{6.09 \times 10^{-7} \ \frac{\text{ft}^2}{\text{sec}}}$$

$$= 0.51 \ \text{ft/sec}$$

$$v_x = Ki = \left(0.51 \ \frac{\text{ft}}{\text{sec}}\right)\left(86{,}400 \ \frac{\text{sec}}{\text{day}}\right)(0.000\,35)$$

$$= 15 \ \text{ft/day}$$

(D) This incorrect solution fails to convert intrinsic permeability from in^2 to ft^2. Other assumptions, definitions, and equations are unchanged from the correct solution.

$$K = \frac{\left(32.2 \ \frac{\text{ft}}{\text{sec}^2}\right)(1.4 \times 10^{-6} \ \text{ft}^2)}{1.537 \times 10^{-5} \ \frac{\text{ft}^2}{\text{sec}}}$$

$$= 2.9 \ \text{ft/sec}$$

$$v_x = \frac{\left(2.9 \ \frac{\text{ft}}{\text{sec}}\right)\left(86{,}400 \ \frac{\text{sec}}{\text{day}}\right)(0.000\,35)}{0.42}$$

$$= 209 \ \text{ft/day} \quad (210 \ \text{ft/day})$$

SOLUTION 84

C	equilibrium concentrations	mg/L
C_a	adsorption capacity	μg/g
K_d	distribution coefficient (slope of the isotherm plot)	cm^3/g

$$K_d = \frac{\Delta C_a}{\Delta C} = \frac{\left(250 \frac{\mu g}{g} - 50 \frac{\mu g}{g}\right)\left(\frac{1 \text{ mg}}{1000 \mu g}\right) \times \left(\frac{1000 \text{ mL}}{L}\right)\left(1 \frac{\text{cm}^3}{\text{mL}}\right)}{6.7 \frac{\text{mg}}{L} - 4.1 \frac{\text{mg}}{L}}$$

$$= 76.9 \text{ cm}^3/g$$

B_d soil bulk density g/cm^3
θ aquifer porosity   –
r_f retardation factor   –

$$r_f = 1 + \frac{B_d K_d}{\theta} = 1 + \frac{\left(1.68 \frac{g}{\text{cm}^3}\right)\left(76.9 \frac{\text{cm}^3}{g}\right)}{0.43}$$

$$= 301$$

v_s chemical velocity
v_{gw} groundwater velocity

$$v_s = \frac{v_{gw}}{r_f} = \frac{v_{gw}}{301}$$

$$= 0.0033 \ v_{gw}$$

The answer is (C).

Why Other Options Are Wrong

(A) This incorrect choice divides by the soil organic carbon fraction when calculating the retardation factor. Other assumptions, definitions, and equations are unchanged from the correct solution.

$$K_d = \frac{\left(250 \frac{\mu g}{g} - 50 \frac{\mu g}{g}\right)\left(\frac{1 \text{ mg}}{1000 \mu g}\right) \times \left(1000 \frac{\text{mL}}{L}\right)\left(1 \frac{\text{cm}^3}{\text{mL}}\right)}{6.7 \frac{\text{mg}}{L} - 4.1 \frac{\text{mg}}{L}}$$

$$= 76.9 \text{ cm}^3/g$$

f_{oc} total organic carbon fraction   –

$$f_{oc} = \left(271 \frac{\text{mg}}{\text{kg}}\right)\left(\frac{1 \text{ kg}}{10^6 \text{ mg}}\right) = 0.000\,271$$

$$r_f = 1 + \frac{B_d K_d}{f_{oc}\theta} = 1 + \frac{\left(1.68 \frac{g}{\text{cm}^3}\right)\left(76.9 \frac{\text{cm}^3}{g}\right)}{(0.43)(0.000\,271)}$$

$$= 1.1 \times 10^6$$

$$v_s = \frac{v_{gw}}{1.1 \times 10^6} = 9.0 \times 10^{-7} \ v_{gw}$$

(B) This incorrect choice misreads the units for C_a as mg/g instead of μg/g. Other assumptions, definitions, and equations are unchanged from the correct solution.

$$K_d = \frac{\left(250 \frac{\text{mg}}{g} - 50 \frac{\text{mg}}{g}\right)\left(1000 \frac{\text{mL}}{L}\right)\left(\frac{1 \text{ cm}^3}{1 \text{ mL}}\right)}{\left(6.7 \frac{\text{mg}}{L} - 4.1 \frac{\text{mg}}{L}\right)}$$

$$= 76\,900 \text{ cm}^3/g$$

$$r_f = 1 + \frac{\left(1.68 \frac{g}{\text{cm}^3}\right)\left(76\,900 \frac{\text{cm}^3}{g}\right)\left(1 \frac{\text{cm}^3}{\text{mL}}\right)}{0.43}$$

$$= 3.0 \times 10^5$$

$$v_s = \frac{v_{gw}}{3.0 \times 10^5} = 3.3 \times 10^{-6} \ v_{gw}$$

(D) This incorrect choice inverts the slope and ignores the units. Other assumptions, definitions, and equations are unchanged from the correct solution.

$$K_d = \frac{\left(6.7 \frac{\text{mg}}{L} - 4.1 \frac{\text{mg}}{L}\right)}{\left(250 \frac{\mu g}{g} - 50 \frac{\mu g}{g}\right)}$$

$$= 0.013 \text{ cm}^3/g$$

Units do not work.

$$r_f = 1 + \frac{\left(1.68 \frac{g}{\text{cm}^3}\right)\left(0.013 \frac{\text{cm}^3}{g}\right)}{0.43} = 1.05$$

$$v_s = \frac{v_{gw}}{1.05} = 0.95 \ v_{gw}$$

SOLUTION 85

Assume 100 mg of total solvent distributed as 79 mg of tetrachloroethene (PCE), 11 mg trichloroethene (TCE), and 10 mg trans-1,2-dichloroethene (t-1,2-DCE).

For PCE, the solubility is 150 mg/L.

The mole weight is

$$(2)\left(12 \frac{\text{mg}}{\text{mmol}}\right) + (4)\left(35.5 \frac{\text{mg}}{\text{mmol}}\right) = 166 \text{ mg/mmol}$$

The mole quantity is

$$\frac{79 \text{ mg}}{166 \frac{\text{mg}}{\text{mmol}}} = 0.48 \text{ mmol}$$

For TCE, the solubility is 1100 mg/L.

The mole weight is

$$(2)\left(12 \frac{\text{mg}}{\text{mmol}}\right) + 1 \frac{\text{mg}}{\text{mmol}} + (3)\left(35.5 \frac{\text{mg}}{\text{mmol}}\right)$$

$$= 131.5 \text{ mg/mmol}$$

The mole quantity is

$$\frac{11 \text{ mg}}{131.5 \frac{\text{mg}}{\text{mmol}}} = 0.084 \text{ mmol}$$

For t-1,2-DCE, the solubility is 600 mg/L.

The mole weight is

$$(2)\left(12 \frac{\text{mg}}{\text{mmol}}\right) + (2)\left(1 \frac{\text{mg}}{\text{mmol}}\right) + (2)\left(35.5 \frac{\text{mg}}{\text{mmol}}\right)$$
$$= 97 \text{ mg/mmol}$$

The mole quantity is

$$\frac{10 \text{ mg}}{97 \frac{\text{mg}}{\text{mmol}}} = 0.10 \text{ mmol}$$

The mole total is

$$0.48 \text{ mmol} + 0.084 \text{ mmol} + 0.10 \text{ mmol} = 0.664 \text{ mmol}$$

C_{PCE}	equilibrium concentration of PCE in the mixture	mg/L
S_{PCE}	water solubility of PCE	mg/L
x_{PCE}	mole fraction of PCE in the mixture	mol/mol

$$C_{PCE} = x_{PCE} S_{PCE}$$
$$= \left(\frac{0.48 \text{ mmol}}{0.664 \text{ mmol}}\right)\left(150 \frac{\text{mg}}{\text{L}}\right)$$
$$= 108 \text{ mg/L} \quad (110 \text{ mg/L})$$

The answer is (B).

Why Other Options Are Wrong

(A) This incorrect choice uses mmol of tetrachloroethene (PCE) for the mole fraction and ignores the influence of the other chemicals. Other assumptions, definitions, and equations are unchanged from the correct solution.

$$S_{PCE} = 150 \text{ mg/L}$$

For PCE, the mole weight is

$$(2)\left(12 \frac{\text{mg}}{\text{mmol}}\right) + (4)\left(35.5 \frac{\text{mg}}{\text{mmol}}\right) = 166 \text{ mg/mmol}$$

The mole quantity is

$$\frac{79 \text{ mg}}{166 \frac{\text{mg}}{\text{mmol}}} = 0.48 \text{ mmol}$$

$$C_{PCE} = (0.48 \text{ mmol})\left(150 \frac{\text{mg}}{\text{L}}\right) = 72 \text{ mg/L}$$

Note that the mmol units are ignored.

(C) This incorrect choice multiplies the water solubility of tetrachloroethene (PCE) by the fraction of PCE in the mixture. Other definitions are unchanged from the correct solution.

$$S_{PCE} = 150 \text{ mg/L}$$
$$C_{PCE} = (0.79)\left(150 \frac{\text{mg}}{\text{L}}\right)$$
$$= 119 \text{ mg/L} \quad (120 \text{ mg/L})$$

(D) This incorrect choice assumes that the solubility of tetrachloroethene (PCE) in water and the equilibrium concentration of PCE in solution with other chemicals are equal. Other definitions are unchanged from the correct solution.

$$S_{PCE} = C_{PCE} = 150 \text{ mg/L}$$

SOLUTION 86

Assume diffusion is negligible for a hydraulic conductivity of 0.42 m/d.

D_L	longitudinal hydrodynamic dispersion	m²/d
i	gradient	m/m
K	hydraulic conductivity	m/d
n_e	effective porosity	–
α_L	dynamic dispersivity	m

$$D_L = \frac{\alpha_L K i}{n_e}$$
$$= \frac{(872 \text{ m})\left(0.42 \frac{\text{m}}{\text{d}}\right)\left(0.0012 \frac{\text{m}}{\text{m}}\right)}{0.36}$$
$$= 1.2 \text{ m}^2/\text{d}$$

The answer is (D).

Why Other Options Are Wrong

(A) This incorrect solution calculates the dynamic dispersivity using 872 m as a distance between wells and uses the Darcy velocity instead of the actual groundwater velocity by failing to divide by the effective porosity. Other assumptions and definitions are unchanged from the correct solution.

$$L = \text{distance between wells} = 872 \text{ m}$$
$$\alpha_L = 0.0175 L^{1.46} = (0.0175)(872 \text{ m})^{1.46}$$
$$= 344 \text{ m}$$
$$D_L = \alpha_L K i = (344 \text{ m})\left(0.42 \frac{\text{m}}{\text{d}}\right)\left(0.0012 \frac{\text{m}}{\text{m}}\right)$$
$$= 0.17 \text{ m}^2/\text{d}$$

(B) This incorrect solution uses the Darcy velocity instead of the actual groundwater velocity by failing to divide by the effective porosity. Other definitions and assumptions are unchanged from the correct solution.

$$D_L = \alpha_L K i = (872 \text{ m})\left(0.42 \, \frac{\text{m}}{\text{d}}\right)\left(0.0012 \, \frac{\text{m}}{\text{m}}\right)$$
$$= 0.44 \text{ m}^2/\text{d}$$

(C) This incorrect solution uses effective porosity as effective diffusion, and uses the Darcy velocity instead of the actual groundwater velocity by failing to divide by the effective porosity. Other definitions are unchanged from the correct solution.

D^* effective diffusion m^2/d

$$D_L = \alpha_L K i + D^*$$
$$= (872 \text{ m})\left(0.42 \, \frac{\text{m}}{\text{d}}\right)\left(0.0012 \, \frac{\text{m}}{\text{m}}\right) + 0.36 \, \frac{\text{m}^2}{\text{d}}$$
$$= 0.80 \text{ m}^2/\text{d}$$

SOLUTION 87

Assume the release is longitudinally one dimensional, continuous, and at steady state, and assume that diffusion is negligible.

L distance between points of interest   m
α_L dispersion coefficient   m

$$\alpha_L = 0.0175 L^{1.46} = (0.0175)(1600 \text{ m})^{1.46} = 834 \text{ m}$$

D_L longitudinal hydrodynamic dispersion m^2/d
v_x bulk groundwater velocity   m/d

$$D_L = \alpha_L v_x = (834 \text{ m})\left(172 \, \frac{\text{cm}}{\text{d}}\right)\left(\frac{1 \text{ m}}{100 \text{ cm}}\right)$$
$$= 1434 \text{ m}^2/\text{d}$$

C concentration of the solute at time t mg/L
C_o concentration of the solute at time zero   mg/L
D_L longitudinal hydrodynamic dispersion m^2/d
erfc   complementary error function   –
t travel time of interest   day

$$\frac{0.5C}{C_o} = \text{erfc} \frac{L - v_x t}{2\sqrt{D_L t}} = \frac{(0.5)\left(0.001 \, \frac{\text{mg}}{\text{L}}\right)}{0.182 \, \frac{\text{mg}}{\text{L}}}$$
$$= 0.00275$$

$$\frac{L - v_x t}{2\sqrt{D_L t}} = 2.1 = \frac{1600 \text{ m} - \left(172 \, \frac{\text{cm}}{\text{d}}\right)\left(\frac{1 \text{ m}}{100 \text{ cm}}\right) t}{2\sqrt{\left(1434 \, \frac{\text{m}^2}{\text{d}}\right) t}}$$

Multiply through and square and combine terms.
$$t^2 - 10411 t + 8.7 \times 10^5 = 0$$

Solve for t using the quadratic formula.
$$t = 84 \text{ d}$$

The answer is (B).

Why Other Options Are Wrong

(A) This incorrect solution fails to convert the groundwater velocity units from cm/d to m/d. Other assumptions, definitions, and equations are the same as used in the correct solution.

$$\alpha_L = 0.0175 L^{1.46} = (0.0175)(1600 \text{ m})^{1.46}$$
$$= 834 \text{ m}$$

$$D_L = \alpha_L v_x = (834 \text{ m})\left(172 \, \frac{\text{cm}}{\text{d}}\right)$$
$$= 143\,448 \text{ m}^2/\text{d}$$

Note that units are incorrect.

$$\frac{0.5C}{C_o} = \text{erfc} \frac{L - v_x t}{2\sqrt{D_L t}} = \frac{(0.5)\left(0.001 \, \frac{\text{mg}}{\text{L}}\right)}{0.182 \, \frac{\text{mg}}{\text{L}}}$$
$$= 0.00275$$

$$\frac{L - v_x t}{2\sqrt{D_L t}} = 2.1 = \frac{1600 \text{ m} - \left(172 \, \frac{\text{cm}}{\text{d}}\right) t}{2\sqrt{\left(143\,448 \, \frac{\text{m}^2}{\text{d}}\right) t}}$$

Note that units are incorrect.

Multiply through and square and combine terms.
$$t^2 - 104 t + 87 = 0$$

Solve for t using the quadratic formula.
$$t = 0.8 \text{ d}$$

(C) This incorrect solution misuses the complementary error function. Other assumptions, definitions, and equations are the same as used in the correct solution.

$$\alpha_L = 0.0175 L^{1.46} = (0.0175)(1600 \text{ m})^{1.46}$$
$$= 834 \text{ m}$$

$$D_L = \alpha_L v_x = (834 \text{ m})\left(172 \, \frac{\text{cm}}{\text{d}}\right)\left(\frac{1 \text{ m}}{100 \text{ cm}}\right)$$
$$= 1434 \text{ m}^2/\text{d}$$

$$\frac{0.5C}{C_o} = \text{erfc} \frac{L - v_x t}{2\sqrt{D_L t}} = \frac{(0.5)\left(0.001 \, \frac{\text{mg}}{\text{L}}\right)}{0.182 \, \frac{\text{mg}}{\text{L}}}$$
$$= 0.00275$$

$$\frac{L - v_x t}{2\sqrt{D_L t}} = 0.0476 = \frac{1600 \text{ m} - \left(172 \, \frac{\text{cm}}{\text{d}}\right)\left(\frac{1 \text{ m}}{100 \text{ cm}}\right) t}{2\sqrt{\left(1434 \, \frac{\text{m}^2}{\text{d}}\right) t}}$$

Multiply through and square and combine terms.
$$t^2 - 1860t + 8.60 \times 10^5 = 0$$
Solve for t using the quadratic formula.
$$t = 860 \text{ d}$$

(D) This incorrect solution misapplies the quadratic formula. Other assumptions, definitions, and equations are the same as used in the correct solution.
$$\alpha_L = 0.0175 L^{1.46}$$
$$= (0.0175)(1600 \text{ m})^{1.46}$$
$$= 834 \text{ m}$$
$$D_L = \alpha_L v_x$$
$$= (834 \text{ m})\left(172 \frac{\text{cm}}{\text{d}}\right)\left(\frac{1 \text{ m}}{100 \text{ cm}}\right)$$
$$= 1434 \text{ m}^2/\text{d}$$
$$\frac{0.5C}{C_o} = \text{erfc}\frac{L - v_x t}{2\sqrt{D_L t}}$$
$$= \frac{(0.5)\left(0.001 \frac{\text{mg}}{\text{L}}\right)}{0.182 \frac{\text{mg}}{\text{L}}}$$
$$= 0.00275$$
$$\frac{L - v_x t}{2\sqrt{D_L t}} = 2.1$$
$$= \frac{1600 \text{ m} - \left(172 \frac{\text{cm}}{\text{d}}\right)\left(\frac{1 \text{ m}}{100 \text{ cm}}\right)t}{2\sqrt{\left(1434 \frac{\text{m}^2}{\text{d}}\right)t}}$$

Multiply through and square and combine terms.
$$t^2 - 10\,411t + 8.7 \times 10^5 = 0$$
Solve for t using the quadratic formula.
$$t = 5280 \text{ d} \quad (5300 \text{ d})$$

SOLUTION 88

d_i	layer thickness	cm
K	overall hydraulic conductivity	cm/s
K_i	layer hydraulic conductivity	cm/s

$$K = \frac{\sum d_i}{\sum \frac{d_i}{K_i}}$$
$$= \frac{130 \text{ cm} + 180 \text{ cm} + 270 \text{ cm} + 65 \text{ cm}}{\frac{130 \text{ cm}}{0.0090 \frac{\text{cm}}{\text{s}}} + \frac{180 \text{ cm}}{0.017 \frac{\text{cm}}{\text{s}}}}$$
$$+ \frac{270 \text{ cm}}{0.036 \frac{\text{cm}}{\text{s}}} + \frac{65 \text{ cm}}{0.011 \frac{\text{cm}}{\text{s}}}$$
$$= 0.0168 \text{ cm/s}$$

A	unit area of aquifer	1 m²
i	hydraulic gradient	m/m, taken as $\Delta h/1$ m
q	vertical ground water flow rate per unit area of aquifer	m³/s·m²

$$q = KiA$$
$$= \left(0.0168 \frac{\text{cm}}{\text{s}}\right)\left(14 \frac{\text{cm}}{\text{m}}\right)\left(\frac{1 \text{ m}^2}{10\,000 \text{ cm}^2}\right)(1 \text{ m}^2)$$
$$= 2.35 \times 10^{-5} \text{ m}^3/\text{s for 1 m}^2 \text{ of aquifer area}$$

The answer is (B).

Why Other Options Are Wrong

(A) This incorrect solution uses the average of the hydraulic conductivities and the ratio of the thicknesses of the five layers to that of the four layers. Other assumptions, definitions, and equations are unchanged from the correct solution.

$$K = \frac{\left(0.0090 \frac{\text{cm}}{\text{s}} + 0.017 \frac{\text{cm}}{\text{s}} + 0.036 \frac{\text{cm}}{\text{s}} + 0.011 \frac{\text{cm}}{\text{s}}\right)}{(4)\left(\frac{130 \text{ cm} + 180 \text{ cm}}{+ 270 \text{ cm} + 65 \text{ cm} + 110 \text{ cm}}\right)} \times (130 \text{ cm} + 180 \text{ cm} + 270 \text{ cm} + 65 \text{ cm})$$
$$= 0.0156 \text{ cm/s}$$
$$q = \left(0.0156 \frac{\text{cm}}{\text{s}}\right)\left(14 \frac{\text{cm}}{\text{m}}\right)\left(\frac{1 \text{ m}^2}{10\,000 \text{ cm}^2}\right)(1 \text{ m}^2)$$
$$= 2.18 \times 10^{-5} \text{ m}^3/\text{s for 1 m}^2 \text{ of aquifer area}$$

(C) This incorrect solution includes all layers in the calculation for overall hydraulic conductivity, not only those layers defining the flow path between the screened intervals of the two wells. Other assumptions, definitions, and equations are unchanged from the correct solution.

$$K = \frac{130 \text{ cm} + 180 \text{ cm} + 270 \text{ cm} + 65 \text{ cm} + 110 \text{ cm}}{\frac{130 \text{ cm}}{0.0090 \frac{\text{cm}}{\text{s}}} + \frac{180 \text{ cm}}{0.017 \frac{\text{cm}}{\text{s}}}}$$
$$+ \frac{270 \text{ cm}}{0.036 \frac{\text{cm}}{\text{s}}} + \frac{65 \text{ cm}}{0.011 \frac{\text{cm}}{\text{s}}} + \frac{110 \text{ cm}}{0.020 \frac{\text{cm}}{\text{s}}}$$
$$= 0.0172 \text{ cm/s}$$
$$q = \left(0.0172 \frac{\text{cm}}{\text{s}}\right)\left(14 \frac{\text{cm}}{\text{m}}\right)\left(\frac{1 \text{ m}^2}{10\,000 \text{ cm}^2}\right)(1 \text{ m}^2)$$
$$= 2.41 \times 10^{-5} \text{ m}^3/\text{s for 1 m}^2 \text{ of aquifer area}$$

(D) This incorrect solution uses the average hydraulic conductivity of the soil layers. Other assumptions, definitions, and equations are unchanged from the correct solution.

$$K = \frac{0.0090 \frac{\text{cm}}{\text{s}} + 0.017 \frac{\text{cm}}{\text{s}} + 0.036 \frac{\text{cm}}{\text{s}} + 0.011 \frac{\text{cm}}{\text{s}}}{4}$$
$$= 0.0183 \text{ cm/s}$$

$$q = \left(0.0183 \frac{\text{cm}}{\text{s}}\right)\left(14 \frac{\text{cm}}{\text{m}}\right)\left(\frac{1 \text{ m}^2}{10\,000 \text{ cm}^2}\right)(1 \text{ m}^2)$$
$$= 2.56 \times 10^{-5} \text{ m}^3/\text{s for 1 m}^2 \text{ of aquifer area}$$

SOLUTION 89

Inorganic chemical contaminants may be either reduced or oxidized to an insoluble form that would demobilize the contaminant as a precipitate. Any oxidizing or reducing reagent employed needs to be completely reactive in that no undesired residual remains. For example, hydrogen peroxide would be a suitable oxidizing reagent since it decomposes to free oxygen and water, products that do not represent contaminants themselves. However, chlorine compounds would not be suitable for in-situ oxidation since trihalomethane (THM) formation and other unwanted byproducts may result. Also, a cut-off wall backfilled with a material such as crushed limestone to facilitate precipitating reactions or activated carbon to adsorb contaminants may be effective if groundwater is relatively shallow. For inorganic chemical contaminants, enhanced biodegradation would typically not be effective since inorganic chemicals do not provide a carbon source for biological activity and are not biologically degradable.

The answer is (C).

Why Other Options Are Wrong

(A) This choice is incorrect because chemical oxidation using hydrogen peroxide may be appropriate for oxidizing inorganic contaminants.

(B) This choice is incorrect because, although chemical oxidation through superchlorination is not an appropriate method for in situ remediation, chemical oxidation using hydrogen peroxide may be appropriate for oxidizing inorganic contaminants.

(D) This choice is incorrect because using a cut-off wall backfilled with reactive or adsorption material would be a potentially suitable method for in situ remediation of inorganic contaminants, especially where groundwater is relatively shallow.

SOLUTION 90

Assume that the contaminant is acetate and that the ammonia concentration is as ammonium (NH_4^+).

The mole weight of NH_4^+ is

$$14 \frac{\text{g}}{\text{mol}} + (4)\left(1 \frac{\text{g}}{\text{mol}}\right) = 18 \text{ g/mol}$$

The mole weight of CH_3COO^- is

$$(2)\left(12 \frac{\text{g}}{\text{mol}}\right) + (3)\left(1 \frac{\text{g}}{\text{mol}}\right) + (2)\left(16 \frac{\text{g}}{\text{mol}}\right)$$
$$= 59 \text{ g/mol}$$

The molar concentration of ammonium is

$$\frac{\left(9.7 \frac{\text{mg NH}_4^+}{\text{L}}\right)\left(\frac{1 \text{ g}}{1000 \text{ mg}}\right)}{18 \frac{\text{g}}{\text{mol}}} = 0.54 \frac{\text{mol NH}_4^+}{\text{L}}$$

Six mol of ammonium will react with 26 mol of acetate. Using this ratio, the mass concentration of acetate is

$$\frac{\left(0.54 \frac{\text{mol NH}_4^+}{\text{L}}\right)(26 \text{ mol CH}_3\text{COO}^-)}{(6 \text{ mol NH}_4^+)} \times \left(59 \frac{\text{g}}{\text{mol}}\right)\left(\frac{1000 \text{ mg}}{\text{g}}\right)$$
$$= 138 \text{ mg/L} \quad (140 \text{ mg/L})$$

The answer is (C).

Why Other Options Are Wrong

(A) This incorrect solution inverts the mole ratios of ammonium and acetate. Other assumptions are the same as used in the correct solution.

The mole weight of ammonium is

$$14 \frac{\text{g}}{\text{mol}} + (4)\left(1 \frac{\text{g}}{\text{mol}}\right) = 18 \text{ g/mol}$$

The mole weight of acetate is

$$(2)\left(12 \frac{\text{g}}{\text{mol}}\right) + (3)\left(1 \frac{\text{g}}{\text{mol}}\right) + (2)\left(16 \frac{\text{g}}{\text{mol}}\right)$$
$$= 59 \text{ g/mol}$$

The molar concentration of ammonium is

$$\frac{\left(9.7 \frac{\text{mg NH}_4^+}{\text{L}}\right)\left(\frac{1 \text{ g}}{1000 \text{ mg}}\right)}{18 \frac{\text{g}}{\text{mol}}} = 0.54 \text{ mol NH}_4^+/\text{L}$$

The mass concentration of acetate is

$$\frac{\left(0.54 \frac{\text{mol NH}_4^+}{\text{L}}\right)(6 \text{ mol CH}_3\text{COO}^-)}{(26 \text{ mol NH}_4^+)} \times \left(59 \frac{\text{g}}{\text{mol}}\right)\left(\frac{1000 \text{ mg}}{\text{g}}\right)$$
$$= 7.4 \text{ mg/L}$$

(B) This incorrect solution calculates the ammonia requirement. Assumptions are the same as used in the correct solution.

The mole weight of NH_4^+ is

$$14\ \frac{g}{mol} + (4)\left(1\ \frac{g}{mol}\right) = 18\ g/mol$$

The mole weight of CH_3COO^- is

$$(2)\left(12\ \frac{g}{mol}\right) + (3)\left(1\ \frac{g}{mol}\right) + (2)\left(16\ \frac{g}{mol}\right)$$
$$= 59\ g/mol$$

The molar concentration of acetate is

$$\frac{\left(818\ \frac{mg\ CH_3COO^-}{L}\right)\left(\frac{1\ g}{1000\ mg}\right)}{59\ \frac{g}{mol}}$$
$$= 13.9\ mol\ CH_3COO^-/L$$

Six mol of ammonium will react with 26 mol of acetate. Using this ratio, the mass concentration of ammonium, incorrectly taken as acetate, is

$$\frac{(6\ mol\ NH_4^+)\left(13.9\ \frac{mol\ CH_3COO^-}{L}\right)}{(26\ mol\ CH_3COO^-)}$$
$$\times \left(18\ \frac{g}{mol}\right)\left(\frac{1000\ mg}{g}\right)$$
$$= 58\ mg/L$$

(D) This incorrect solution mistakenly uses the biomass as the contaminant and calculates the biomass produced by the chemical biodegradation. Other assumptions are the same as used in the correct solution.

Assume $C_5H_7O_2N$ is the contaminant.

The mole weight of $C_5H_7O_2N$ is

$$(5)\left(12\ \frac{g}{mol}\right) + (7)\left(1\ \frac{g}{mol}\right) + (2)\left(16\ \frac{g}{mol}\right)$$
$$+ 14\ \frac{g}{mol}$$
$$= 113\ g/mol$$

The molar concentration of contaminant is

$$\frac{\left(818\ \frac{mg\ C_5H_7O_2N}{L}\right)\left(\frac{1\ g}{1000\ mg}\right)}{113\ \frac{g}{mol}}$$
$$= 7.2\ mol\ C_5H_7O_2N/L$$

The mass concentration of contaminant is

$$\left(7.2\ \frac{mol\ C_5H_7O_2N}{L}\right)\left(113\ \frac{g}{mol}\right)\left(\frac{1000\ mg}{g}\right)$$
$$= 814\ mg/L\ \ (810\ mg/L)$$

SOLUTION 91

The flow lines illustrate the path along which an ideal water particle travels on its way to the extraction wells pumping at steady state. The tick marks illustrate the distance traveled by an ideal water particle in uniform time increments. In the illustration, the time increment is identified as 10 days. Where the tick marks are spaced very closely, the water travels a very short distance during the 10 day time increment. In the area between two extraction wells, the water particle is being acted upon by each well and effectively becomes suspended between the two. This represents stagnation and is shown in the illustration at A. At B and C the tick mark spacing is great enough to show that the water is flowing toward the wells—not indicative of stagnation. D indicates an extraction well, the point of highest groundwater velocity.

The answer is (A).

Why Other Options Are Wrong

(B) This is an incorrect option because B identifies a point in the outer regions of the well capture zone where water is moving toward the wells and is far distant from the stagnation zones located between adjacent extraction wells.

(C) This is an incorrect option because C identifies a point located near an extraction well where water is moving along the flow path directly toward the well. The flow lines and tick marks in this area are not illustrative of stagnation.

(D) This is an incorrect option because D identifies one of the extraction wells. At the well, the water is moving at its maximum velocity, a condition effectively opposite that of stagnation.

SOLUTION 92

K_La	mass-transfer coefficient	s^{-1}
NTU	number of transfer units	–
R_f	recovered fraction	–
S	stripping factor	–

$$NTU = \left(\frac{S}{S-1}\right)\ln\frac{(1-R_f)^{-1}(S-1)+1}{S}$$
$$= \left(\frac{3}{3-1}\right)\ln\frac{(1-0.996)^{-1}(3-1)+1}{3}$$
$$= 7.68$$

HLR	volumetric flow rate	$m^3/m^2 \cdot s$
HTU	transfer unit height	m

$$\text{HTU} = \frac{\text{HLR}}{K_L a} = \frac{0.022 \, \frac{\text{m}^3}{\text{m}^2 \cdot \text{s}}}{\frac{0.013}{\text{s}}}$$

$$= 1.69 \text{ m}$$

The packing height is

$$(\text{HTU})(\text{NTU}) = (1.69 \text{ m})(7.68)$$
$$= 13 \text{ m}$$

The answer is (B).

Why Other Options Are Wrong

(A) This incorrect solution inverts the equation for HTU. Other definitions and equations are unchanged from the correct solution.

$$\text{NTU} = \left(\frac{3}{3-1}\right) \ln \left(\frac{(1-0.966)^{-1}(3-1)+1}{3}\right)$$
$$= 7.68$$

$$\text{HTU} = \frac{K_L a}{\text{HLR}} = \frac{\frac{0.013}{\text{s}}}{0.022 \, \frac{\text{m}^3}{\text{m}^2 \cdot \text{s}}} = 0.59 \text{ m}$$

Units do not work.

The packing height is

$$(\text{HTU})(\text{NTU}) = (0.59 \text{ m})(7.68)$$
$$= 4.5 \text{ m}$$

(C) This incorrect solution improperly inverts the removal fraction term in the NTU equation. Other definitions and equations are unchanged from the correct solution.

$$\text{NTU} = \left(\frac{S}{S-1}\right) \ln \left(\frac{(S-1)+1}{(1-R_f)S}\right)$$
$$= \left(\frac{3}{3-1}\right) \ln \left(\frac{(3-1)+1}{(1-0.996)(3)}\right)$$
$$= 8.28$$

$$\text{HTU} = \frac{0.022 \, \frac{\text{m}^3}{\text{m}^2 \cdot \text{s}}}{\frac{0.013}{\text{s}}}$$
$$= 1.69 \text{ m}$$

The packing height is

$$(\text{HTU})(\text{NTU}) = (1.69 \text{ m})(8.28)$$
$$= 14 \text{ m}$$

(D) This incorrect solution misinterprets the NTU equation. Other definitions and equations are unchanged from the correct solution.

$$\text{NTU} = \left(\frac{S}{S-1}\right) \ln \left((1-R_f)^{-1}(S-1) + \frac{1}{S}\right)$$
$$= \left(\frac{3}{3-1}\right) \ln \left((1-0.996)^{-1}(3-1) + \frac{1}{3}\right)$$
$$= 9.32$$

$$\text{HTU} = \frac{0.022 \, \frac{\text{m}^3}{\text{m}^2 \cdot \text{s}}}{\frac{0.013}{\text{s}}} = 1.69 \text{ m}$$

The packing height is

$$(\text{HTU})(\text{NTU}) = (1.69 \text{ m})(9.32) = 16 \text{ m}$$

SOLUTION 93

Gasoline that has leaked from an underground tank and later been discovered in a monitoring well will experience weathering. Weathering represents a loss of lighter fractions to evaporation and other physical and chemical changes that creates on the chromatograph an increase in the proportion of heavier fractions.

The lighter fractions occur earlier and the heavier fractions occur later on the chromatograph. For a given gasoline, the timing of the peaks will be the same between fresh and weathered gasoline, but their relative magnitude will change.

Comparing the choices with the illustration given in the problem statement, the illustration at (A) is identical and does not represent any changes due to weathering or other phenomena. The illustration at (B) shows peaks occurring at different times than the problem statement illustration, indicating a different gasoline. The illustration at (D) is enriched in lighter fractions, showing the opposite affect of weathering. Only the illustration at (C) shows the peaks occurring at the same time as the problem statement illustration with enrichment in the heavier fractions.

The answer is (C).

Why Other Options Are Wrong

(A) The illustration is identical to the problem statement illustration and does not show any changes due to weathering or other phenomena. Weathering would have occurred between the time of the release from the underground tank and its discovery in the monitoring well.

(B) The illustration shows peaks occurring at different times than those of the problem statement illustration. This indicates that the gasolines represented by the two illustrations are not the same and do not have the underground tank as their common origin.

(D) The illustration shows enrichment in lighter fractions, an unexpected phenomenon that would be inconsistent with a gasoline release. Instead of a relative increase in lighter fractions, a relative increase in heavier fractions would be expected.

SOLUTION 94

Dispersion and diffusion cause dilution of the solute. In more permeable materials, dispersion is far more significant. Diffusion becomes more important in less permeable soils and may occur even where groundwater flow is zero.

Dispersion is caused by mixing in the pore spaces and by differential flow in flow channels, a phenomenon that becomes more significant as the gradient and permeability of the soil increase. Where dispersion dominates, the result is a more pronounced elongated plume that moves away from the source down the gradient.

Diffusion accounts for the molecular spreading of the solute across the boundary between relatively contaminated water and clean water. The fluid does not have to be moving for diffusion to occur. In more permeable media, diffusion is most pronounced in its impact at the boundary. It tends to flatten the concentration gradient across the boundary. In low permeability soils, however, diffusion may create a radial plume around the source.

For the conditions described in the problem statement, diffusion will likely dominate and the plume will be relatively circular in shape.

The answer is (D).

Why Other Options Are Wrong

(A) This choice is incorrect because dispersion dominates under conditions of steeper gradient and higher permeability. In a low permeability soil with very shallow gradient, dispersion will less significantly influence a solute than will diffusion.

(B) This choice is incorrect because dispersion dominates under conditions of steeper gradient and higher permeability. These are the opposite of the conditions given in the problem statement.

(C) This choice is incorrect because, although diffusion dominates under the conditions described in the problem statement, the plume will expand radially around the source and not in a pronounced elongated shape.

SOLUTION 95

As nonaqueous phase liquid (NAPL) moves through the soil, whether in the saturated or unsaturated zones, residual NAPL is left behind. Residual saturation occurs because as the NAPL migrates through the soil, the pressure forces are adequate to force the NAPL into pore spaces, but the gravity forces are not adequate to drain them. This becomes more pronounced as the NAPL viscosity increases and soil permeability decreases.

Because it is pre-wetted by the presence of water, the saturated zone typically experiences a lower residual saturation than does the unsaturated zone. Also, a fluctuating water table will either dislodge the NAPL, causing it to migrate, or not allow the NAPL to enter the pore space. In either case, a lower residual saturation will result.

Of the choices given, only an increase in the percentage of fines in the soil will contribute to an increase in residual saturation because increasing fines reduces the individual pore volume, causing a wider dispersion of the NAPL and making the pores less susceptible to gravity drainage.

The answer is (C).

Why Other Options Are Wrong

(A) This choice is incorrect because pre-wetting the soil will reduce the potential residual saturation. In pre-wetted soils, the nonaqueous phase liquid (NAPL) is not sorbed by the soil surface and loses a portion of the pore volume to the wetting fluid.

(B) This choice is incorrect because pre-wetting and dramatic water table fluctuations both contribute to reduced pore volume for the nonaqueous phase liquid (NAPL) to occupy. Also, the fluctuating water table will act to dislodge some of the NAPL and increase its potential for broader dispersion. Both of these conditions contribute to a lower residual saturation.

(D) This choice is incorrect because, although increased fines in the soil will contribute to an increased residual saturation, a decreased nonaqueous phase liquid (NAPL) viscosity will not. As the NAPL viscosity is decreased, it flows more easily and will be more susceptible to gravity drainage from the soil pore spaces.

SOLUTION 96

d_w	well casing diameter	ft
h, h_o	height of drawdown curve measured from the top of confined aquifer at radial distance $r + r_w$ and $r_o + r_w$ from the well	ft
K	hydraulic conductivity	ft/day
m	thickness of the confined aquifer	ft
Q	discharge flow rate	ft³/sec
r, r_o	radial distances from the well casing	ft
r_w	well casing radius	ft

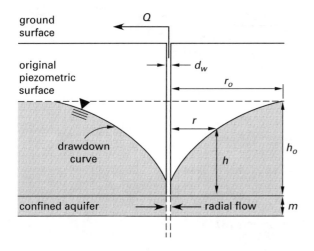

$$r_w = \frac{d_w}{2} = \frac{1 \text{ ft}}{2} = 0.5 \text{ ft}$$

$$Q = \frac{2\pi K m (h_o - h)}{\ln \dfrac{r_o + r_w}{r + r_w}}$$

$$= \frac{2\pi \left(7.3 \dfrac{\text{ft}}{\text{day}}\right) (20 \text{ ft})(105 \text{ ft} - 70 \text{ ft}) \times \left(\dfrac{1 \text{ day}}{86{,}400 \text{ sec}}\right)}{\ln \left(\dfrac{120 \text{ ft} + 0.5 \text{ ft}}{50 \text{ ft} + 0.5 \text{ ft}}\right)}$$

$$= 0.43 \text{ ft}^3/\text{sec}$$

The answer is (C).

Why Other Options Are Wrong

(A) This incorrect solution improperly applies the discharge flow equation. The illustration and other definitions and equations used in the correct solution are unchanged.

$$r_w = \frac{d_w}{2} = \frac{1 \text{ ft}}{2} = 0.5 \text{ ft}$$

$$Q = 2\pi K m \ln \left(\frac{(h_o - h)(r + r_w)}{r_o + r_w}\right)$$

$$= (2)(\pi) \left(7.3 \dfrac{\text{ft}}{\text{day}}\right) (20 \text{ ft}) \left(\dfrac{1 \text{ day}}{86{,}400 \text{ sec}}\right)$$

$$\times \ln \left(\frac{(105 \text{ ft} - 70 \text{ ft})(50 \text{ ft} + 0.5 \text{ ft})}{120 \text{ ft} + 0.5 \text{ ft}}\right)$$

$$= 0.029 \text{ ft}^3/\text{sec}$$

Units do not work.

(B) This incorrect solution neglects to include the natural log function in the discharge equation. The illustration and other definitions and equations used in the correct solution are unchanged.

$$r_w = \frac{d_w}{2} = \frac{1 \text{ ft}}{2} = 0.5 \text{ ft}$$

$$Q = \frac{2\pi K m (h_o - h)(r + r_w)}{r_o + r_w}$$

$$= \frac{(2)(\pi) \left(7.3 \dfrac{\text{ft}}{\text{day}}\right) (20 \text{ ft})(105 \text{ ft} - 70 \text{ ft})}{(120 \text{ ft} + 0.5 \text{ ft}) \left(86{,}400 \dfrac{\text{sec}}{\text{day}}\right)} \times (50 \text{ ft} + 0.5 \text{ ft})$$

$$= 0.16 \text{ ft}^3/\text{sec}$$

(D) This incorrect solution uses the discharge flow equation for an unconfined aquifer instead of the equation for a confined aquifer. Other definitions used in the correct solution are unchanged.

$$r_w = \frac{d_w}{2} = \frac{1 \text{ ft}}{2} = 0.5 \text{ ft}$$

$$Q = \frac{\pi K (h_o^2 - h^2)}{\ln \dfrac{r_o + r_w}{r + r_w}}$$

$$= \frac{\pi \left(7.3 \dfrac{\text{ft}}{\text{day}}\right) ((105 \text{ ft})^2 - (70 \text{ ft}^2)) \left(\dfrac{1 \text{ day}}{86{,}400 \text{ sec}}\right)}{\ln \left(\dfrac{120 \text{ ft} + 0.5 \text{ ft}}{50 \text{ ft} + 0.5 \text{ ft}}\right)}$$

$$= 1.9 \text{ ft}^3/\text{sec}$$

SOLUTION 97

The volume of water lost to storage is the product of the drained aquifer volume and the aquifer storativity.

A_a	horizontal surface area of the aquifer	m²
S	storativity	–
V_d	water volume lost from storage	m³
Δh	change in water table elevation	m

$$V_d = \Delta h S A_a$$

$$= (4.4 \text{ m})(0.21)(512 \text{ km}^2) \left(\frac{1000 \text{ m}}{1 \text{ km}}\right)^2$$

$$= 473{,}088{,}000 \text{ m}^3 \quad (4.7 \times 10^8 \text{ m}^3)$$

The answer is (B).

Why Other Options Are Wrong

(A) This incorrect solution uses the hydraulic conductivity, drained area thickness, and monitoring period to calculate volume. Other definitions are unchanged from the correct solution.

Assume 30 days per month.

K	hydraulic conductivity	m/d
t	monitoring period	d

$$V_d = \Delta h K t = (4.4 \text{ m})\left(0.38 \frac{\text{cm}}{\text{s}}\right)\left(\frac{1 \text{ m}}{100 \text{ cm}}\right)$$
$$\times \left(86\,400 \frac{\text{s}}{\text{d}}\right)(48 \text{ mo})\left(30 \frac{\text{d}}{\text{mo}}\right)$$
$$= 2\,080\,235 \text{ m}^3 \quad (2.1 \times 10^6 \text{ m}^3)$$

Units do not work.

(C) This incorrect solution uses the product of the drained aquifer volume and the average porosity. Other definitions are unchanged from the correct solution.

n_e average porosity   –

$$V_d = \Delta h n_e A_a$$
$$= (4.4 \text{ m})(0.43)(512 \text{ km}^2)\left(1000 \frac{\text{m}}{\text{km}}\right)^2$$
$$= 968\,704\,000 \text{ m}^3 \quad (9.7 \times 10^8 \text{ m}^3)$$

(D) This incorrect solution calculates the volume of water remaining in the aquifer instead of the water volume lost. Other definitions are unchanged from the correct solution.

h thickness of drained area   m
h_o aquifer thickness   m

$$V_d = (h_o - h)SA_a$$
$$= (38 \text{ m} - 4.4 \text{ m})(0.21)(512 \text{ km}^2)\left(1000 \frac{\text{m}}{\text{km}}\right)^2$$
$$= 3\,612\,672\,000 \text{ m}^3 \quad (3.6 \times 10^9 \text{ m}^3)$$

SOLUTION 98

d distance between bottom of drains and top of impermeable layer   m
h water depth in the drain   m
H vertical distance between the bottom of the drain and the highest point of the piezometric surface   m
K hydraulic conductivity   cm/s
q infiltration rate   cm/s
S drain spacing   m

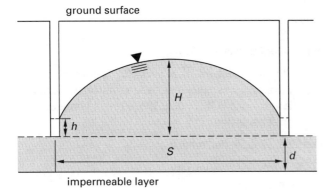

$$S^2 = \frac{4K(H^2 - h^2 + 2dH - 2dh)}{q}$$
$$= \frac{(4)\left(0.018 \frac{\text{cm}}{\text{s}}\right)\left((1.8 \text{ m})^2 - (0.25 \text{ m})^2 + (2)(0.4 \text{ m})(1.8 \text{ m}) - (2)(0.4 \text{ m})(0.25 \text{ m})\right)}{0.0014 \frac{\text{cm}}{\text{s}}}$$
$$= 227 \text{ m}^2$$
$$S = \sqrt{227 \text{ m}^2} = 15 \text{ m}$$

The answer is (B).

Why Other Options Are Wrong

(A) This incorrect solution reverses the values for the hydraulic conductivity (K) and infiltration rate (q) in the spacing equation. The illustration and other definitions are unchanged from the correct solution.

$$S^2 = \frac{4q(H^2 - d^2 + 2hH - 2dh)}{K}$$
$$= \frac{(4)\left(0.0014 \frac{\text{cm}}{\text{s}}\right)\left((1.8 \text{ m})^2 - (0.25 \text{ m})^2 + (2)(0.4 \text{ m})(1.8 \text{ m}) - (2)(0.4 \text{ m})(0.25 \text{ m})\right)}{0.018 \frac{\text{cm}}{\text{s}}}$$
$$= 1.4 \text{ m}^2$$
$$S = \sqrt{1.4 \text{ m}^2} = 1.2 \text{ m}$$

(C) This incorrect solution fails to square the drain spacing (S), vertical distance (H) and water depth (h) terms. The illustration and other definitions are unchanged from the correct solution.

$$S = \frac{4K(H - h + 2dH - 2dh)}{q}$$
$$= \frac{(4)\left(0.018 \frac{\text{cm}}{\text{s}}\right)(1.8 \text{ m} - 0.25 \text{ m} + (2)(0.4 \text{ m})(1.8 \text{ m}) - (2)(0.4 \text{ m})(0.25 \text{ m}))}{0.0014 \frac{\text{cm}}{\text{s}}}$$
$$= 143 \text{ m} \quad (140 \text{ m})$$

(D) This incorrect solution fails to take the square root of the S^2 term. The illustration and other definitions are unchanged from the correct solution.

$$S = \frac{4K(H^2 - h^2 + 2dH - 2dh)}{q}$$

$$= \frac{(4)\left(0.018 \, \frac{\text{cm}}{\text{s}}\right)\left((1.8 \text{ m})^2 - (0.25 \text{ m})^2 \right.}{0.0014 \, \frac{\text{cm}}{\text{s}}}$$
$$\frac{\left. + (2)(0.4 \text{ m})(1.8 \text{ m}) - (2)(0.4 \text{ m})(0.25 \text{ m})\right)}{}$$

$$= 227 \text{ m} \quad (230 \text{ m})$$

Units do not work.

SOLUTION 99

Assume that the auger holes are of equal depth.

a	auger hole diameter	ft
b	distance between auger holes	ft
K	hydraulic conductivity	ft/day
L	aquifer thickness	ft
Q	pumping rate	ft³/day
Δh	head difference between auger holes	ft

$$K = \frac{Q}{\pi L \Delta h} \cosh^{-1} \frac{b}{2a}$$

$$= \left(\frac{\left(26 \, \frac{\text{gal}}{\text{min}}\right)\left(0.134 \, \frac{\text{ft}^3}{\text{gal}}\right)\left(1440 \, \frac{\text{min}}{\text{day}}\right)}{\pi(15 \text{ ft})(9.7 \text{ in})\left(\frac{1 \text{ ft}}{12 \text{ in}}\right)} \right)$$

$$\times \cosh^{-1} \frac{10 \text{ ft}}{(2)(8 \text{ in})\left(\frac{1 \text{ ft}}{12 \text{ in}}\right)}$$

$$= 356 \text{ ft/day} \quad (360 \text{ ft/day})$$

The answer is (C).

Why Other Options Are Wrong

(A) This incorrect solution fails to use consistent units by not converting inches to feet. Other definitions and equations are the same as used in the correct solution.

$$K = \left(\frac{\left(26 \, \frac{\text{gal}}{\text{min}}\right)\left(0.134 \, \frac{\text{ft}^3}{\text{gal}}\right)\left(1440 \, \frac{\text{min}}{\text{day}}\right)}{\pi(15 \text{ ft})(9.7 \text{ in})} \right)$$

$$\times \cosh^{-1} \frac{10 \text{ ft}}{(2)(8 \text{ in})\left(\frac{1 \text{ ft}}{12 \text{ in}}\right)}$$

$$= 30 \text{ ft/day}$$

(B) This incorrect solution uses the cos instead of the $\cosh^{-1}$. Other definitions and equations are the same as used in the correct solution.

$$K = \frac{Q}{\pi L \Delta h} \cos \frac{b}{2a}$$

$$= \left(\frac{\left(26 \, \frac{\text{gal}}{\text{min}}\right)\left(0.134 \, \frac{\text{ft}^3}{\text{gal}}\right)\left(1440 \, \frac{\text{min}}{\text{day}}\right)}{\pi(15 \text{ ft})(9.7 \text{ in})\left(\frac{1 \text{ ft}}{12 \text{ in}}\right)} \right)$$

$$\times \cos \frac{10 \text{ ft}}{(2)(8 \text{ in})\left(\frac{1 \text{ ft}}{12 \text{ in}}\right)}$$

$$= 131 \text{ ft/day} \quad (130 \text{ ft/day})$$

(C) This incorrect solution confuses the distance between wells and the aquifer depth. Other definitions and equations are the same as used in the correct solution.

$$K = \frac{Q}{\pi b \Delta h} \cosh^{-1} \frac{L}{2a}$$

$$= \left(\frac{\left(26 \, \frac{\text{gal}}{\text{min}}\right)\left(0.134 \, \frac{\text{ft}^3}{\text{gal}}\right)\left(1440 \, \frac{\text{min}}{\text{day}}\right)}{\pi(10 \text{ ft})(9.7 \text{ in})\left(\frac{1 \text{ ft}}{12 \text{ in}}\right)} \right)$$

$$\times \cosh^{-1} \frac{15 \text{ ft}}{(2)(8 \text{ in})\left(\frac{1 \text{ ft}}{12 \text{ in}}\right)}$$

$$= 615 \text{ ft/day} \quad (620 \text{ ft/day})$$

SOLUTION 100

Dense nonaqueous phase liquids (DNAPL) are characterized by specific gravities greater than 1.0 and low solubility in water. Of the compounds listed in the table, methylene chloride, naphthalene, and tetrachloroethene satisfy the specific gravity criteria. Of these three, only naphthalene and tetrachloroethene have low water solubilities. Although it may be possible for methylene chloride to exist as a nonaqueous phase liquid, it would only do so under extreme contamination conditions with very high chemical concentrations. Therefore, the most reasonable answer is naphthalene and tetrachloroethene.

The answer is (D).

Why Other Options Are Wrong

(A) This choice is incorrect since acetone will never exist as a dense nonaqueous phase liquid due to its infinite solubility in water.

(B) This choice is incorrect since both methyl ethyl ketone and vinyl chloride have specific gravities less than 1.0. Neither will ever exist as a dense nonaqueous phase liquid.

(C) This choice is incorrect since methylene chloride has a relatively high water solubility that would need to be exceeded before the chemical would exist as a non-aqueous phase liquid. It would be unlikely to encounter methylene chloride at a sufficiently high concentration for it to exist as a nonaqueous phase liquid.

Visit www.ppi2pass.com today to order these essential books for the Civil PE Exam!

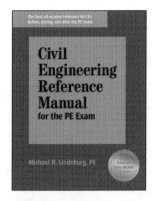

Civil Engineering Reference Manual for the PE Exam
Michael R. Lindeburg, PE

The **Civil Engineering Reference Manual** is the most complete study guide available for engineers preparing for the civil PE exam. It provides a clear, complete review of exam topics, reinforcing key concepts with almost 500 example problems. The text is enhanced by hundreds of illustrations, tables, charts, and formulas, along with a detailed index. After you pass the PE exam, the **Reference Manual** will continue to serve you as a comprehensive desk reference throughout your professional career.

Practice Problems for the Civil Engineering PE Exam: A Companion to the Civil Engineering Reference Manual
Michael R. Lindeburg, PE

The 439 problems in **Practice Problems for the Civil Engineering PE Exam** correspond to chapters in the **Civil Engineering Reference Manual**, giving you problem-solving practice in each topic as you study. Many problems are in the same multiple-choice format as the exam. Complete, step-by-step solutions give you immediate feedback.

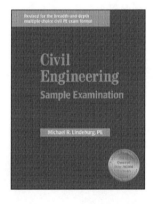

Civil Engineering Sample Examination
Michael R. Lindeburg, PE

Practicing with a realistic simulation of the PE exam is an essential part of your preparation. There's no better way to get ready for the time pressure of the exam. **Civil Engineering Sample Examination** includes a complete eight-hour sample exam, reflecting the exam's breadth-and-depth, multiple-choice format. Full solutions are included.

Six-Minute Solutions for Civil PE Exam Problems
(Available for Environmental, Geotechnical, Structural, Transportation, and Water Resources modules.)

The **Six-Minute Solutions** books help you prepare for the discipline-specific topics of the civil PE exam. Each offers 100 multiple-choice problems providing targeted practice for a particular topic. The 20 morning (breadth) and 80 afternoon (depth) problems are designed to be solved in six minutes—the average amount of time you'll have during the exam. Solutions are included.

For everything you need to pass the exams, go to
www.ppi2pass.com
where you'll find the latest exam news, test-taker advice, the Exam Forum, and secure online ordering.

Professional Publications, Inc.
1250 Fifth Avenue • Belmont, CA 94002
(800) 426-1178 • Fax (650) 592-4519

Source Code BOC ☞ **Quick — *I need additional PPI study materials!***

Please send me the PPI exam review products checked below. I have provided my credit card number and authorize you to charge your current prices, plus shipping, to my card.

For the FE Exam
☐ FE Review Manual
☐ Civil Discipline-Specific Review for the FE/EIT Exam
☐ Mechanical Discipline-Specific Review for the FE/EIT Exam
☐ Electrical Discipline-Specific Review for the FE/EIT Exam
☐ Chemical Discipline-Specific Review for the FE/EIT Exam
☐ Industrial Discipline-Specific Review for the FE/EIT Exam
☐ Engineer-In-Training Reference Manual   ☐ Solutions Manual, SI Units

For the PE, SE, and PLS Exam
☐ Civil Engrg Reference Manual   ☐ Practice Problems
☐ Mechanical Engrg Reference Manual   ☐ Practice Problems
☐ Electrical Engrg Reference Manual   ☐ Practice Problems
☐ Environmental Engrg Reference Manual   ☐ Practice Problems
☐ Chemical Engrg Reference Manual   ☐ Practice Problems
☐ Structural Engrg Reference Manual
☐ PLS Sample Exam

These are just a few of the products we offer for the FE, PE, SE, and LS exams. For a full list, visit our website at www.ppi2pass.com.

NAME/COMPANY _____
STREET _____ SUITE/APT _____
CITY _____ STATE _____ ZIP _____
DAYTIME PH # _____ EMAIL _____
VISA/MC/DSCVR # _____ EXP. DATE _____
CARDHOLDER'S NAME _____
SIGNATURE _____

For fastest service,
Web **www.ppi2pass.com**
Call **800-426-1178**   Fax **650-592-4519**

Mail this form to:
PPI, 1250 Fifth Ave., Belmont, CA 94002

Email Updates Keep You on Top of Your Exam

You need current information to be fully prepared for your exam. Register for PPI's Email Updates to receive convenient updates relevant to the specific exam you are taking. Our updates include notices of exam changes, useful exam tips, errata postings, and new product announcements. There is no charge for this service, and you can cancel at any time.

Register at **www.ppi2pass.com/cgi-bin/signup.cgi**

Free Catalog of Tried-and-True Exam Products

Get a free PPI catalog with a comprehensive selection of the best FE, PE, SE, FLS, and PLS exam-review products available, user tested by more than 800,000 engineers and surveyors. Included are books, software, videos, and the NCEES sample-question books.

Request a catalog at **www.ppi2pass.com/catalogrequest**

How to Report Errors

Find an error? You can report it in two easy steps.

First, check the errata listings on our website, at **www.ppi2pass.com/errata**. The item you noticed may already have been identified. It's always a good idea to check this page before you start studying and periodically thereafter.

Then, go to PPI's Errata Report Form at **www.ppi2pass.com/erratasubmit**, and tell us about the discrepancy you think you've found. Your information will be forwarded to the appropriate author or subject matter expert for verification. Valid corrections are added to the errata section of our website.

You may also fax errata to us at 650-592-4519 or mail them to Professional Publications, Inc., c/o Editorial Errata Department, 1250 Fifth Ave., Belmont, CA 94002.
Be sure to include your name, the book title, the edition and printing numbers, the page number(s), and any other information that will help us locate the error(s).

PROFESSIONAL PUBLICATIONS, INC.
Fax 650-592-4519
errata@ppi2pass.com